DODGING THE STEAMROLLER

A Survival Guide for Century 21

_____

Colin Mason

The Ragged Edge
Publishing

Copyright
© 2016
Colin Mason

'You must dodge the steamroller.'
Bernard Wolfe,
Limbo, 1952

CONTENTS

PART ONE –THE HAZARDS

PART TWO – THE THINGS WE DO

PART ONE – THE HAZARDS

Introduction

By 2030 human civilisation seems likely to face hazards of unprecedented severity. Dodging these, getting by, perhaps even surviving, will depend on how we handle these. Scores, perhaps hundreds of inland cities and towns may be abandoned as summer temperatures soar into the 50s. Huge areas of fertile farmland could degrade into barren desert. Salt-laden seawater will start to flood into the densely settled but low-lying deltas of the great rivers — almost half of the world's food is grown on these. More violent storms, wild fires, destructive rains, will destroy even more of the rapidly diminishing crop yields as well as thousands of flimsy village houses. These things would kill millions, causing destruction, famine and starvation on a huge scale. Those surviving, even in the developed world, would be forced back to life on a medieval scale.

Or

We come to our senses. Fossil fuel use is the major driver of destructive climate change that threatens to visit all the above things on us. We must phase it out within a decade — this would mean a reduction of ten per cent every year from now on. Energy use and all of our transport forms need to be converted to electric power from sustainable sources. We must stop using grains to make motor fuel and to feed meat-producing animals. Every nation, every individual, must address this unprecedented crisis with a degree of urgency generally associated with war. Things like massive defence budgets, regional conflict, unnecessary luxury goods, planned obsolescence and excessive packaging, must somehow be eliminated so the money and technology wasted on them can drive this transformation.

As I hope this book will demonstrate, we really have no choice.

It is based on the idea — axiomatic really — that while bad things are, of course, bad in themselves, much worse can happen if they are allowed to combine. Among these hazards are sea level rise, savage droughts, flood rains, declining food and water resources while populations rise, increasing poverty and displacement of millions of people. All present individual risks, but these will be aggravated if they hit us together — I am calling this only-too-likely convergence the Steamroller.

Once stated, all this may seem obvious, but it pays to look closer, because the Steamroller is not far off, already dangerous and breeding spasms of chaos that are blotting the world like a contagion – infecting Syria, Iraq, Libya, Saudi Arabia, Iran, Chad, Yemen, the Congo, and among its manifestations are the Taliban, Boko Haram in Nigeria, the killings of millions in Rwanda and Cambodia, and the rise of daesh in the Middle East. I will use this admittedly derogatory term throughout this book for the self-styled Islamic state, which is not a state at all but a violent and cruel terror group that world Islam has renounced.

Everything has a cause — these huge underlying hazards, already tending to combine, are prompting social breakdowns that cause an unacceptable toll of death and human suffering. If this is even partly true, much greater efforts to deal with those root causes — the hazards themselves — might be the way to divert, delay, and perhaps even stop the Steamroller in its tracks. For instance, bombing Iraq and Syria to fight daesh destroys people's homes and businesses – what could be more eloquent than pictures of the shattered border city of Kobane? Maybe the effort and money could be better spent improving conditions on the ground, helping communities regain a sane, peaceful way of life, seeing children get a proper schooling, using military force to insulate people from conflict in 'safe zones'?

Most of us know a little about some of the impending hazards, others are less obvious. Can they be dodged or diverted, maybe even stopped? The first step is to establish the truth about these things, and that means recognizing the massive volume of spin, misinformation, crafted to obscure them, and to promote instead what 'they' think you should know. We get this from the climate change deniers who're paid off by big fossil fuel outfits — look carefully and you'll find mining engineers, statisticians, politicians, but precious few climate scientists among the deniers. Then there's the nuclear industry, desperately wanting us to forget about the dire ongoing problems at Chernobyl and Fukushima, which seem likely to persist for hundreds if not thousands of years, because nobody knows what to do about them. And what about those expensive counter-productive wars, Iraq, Afghanistan, costed in the trillions – remember all that about weapons of mass destruction? Before you believe what a 'think-tank' – even one with an impressive-sounding name – is saying, check who is financing it.

So reliable information tends to be in short supply in our fast-changing world, and has to be separated out with some effort from the spin that makes up so much of the current information overload. In the following pages I have tried to do this as simply and briefly as possible, together with some comments on what the real facts imply, what we might do about them. Once we know those facts, and use our intelligence and a resolve to take the necessary action, the Steamroller might be dodged. The prospect then is hopeful, not despairing.

The world is changing, and quickly — Century 21 will be a crunch time, not for the planet, which will survive, but for humanity. An assumption seems to persist that it's the destiny of man to conquer nature, even though it's increasingly plain that nature can and does strike back. Like past species now extinct, we are at odds with her in many ways, and fixing this is urgent, perhaps indeed a matter of survival. Looking at some of the hundreds of books about the various crises that beset Century 21, I was struck first by the amount of detail about what could happen, but most of all by the relative shortage of ideas about what we might do in these hard places, even though rational solutions to at least some are fairly obvious and achievable. These solutions will nevertheless be difficult and expensive because of the deadweight of the business and political establishment, and may require scrapping some of our most cherished concepts – think for instance, nation-state.

There is no shortage of doomsayers predicting the human race will soon become extinct, or at best, that just a few of us will be shunted back into another stone age. Many people, some quite famous, believe we are on track for universal destruction. Leading microbiologist Frank Fenner, who played a major part in the elimination of smallpox, was one of these, saying just before his death in 2010 that 'due to the population explosion, unbridled consumption, and international failure to curb greenhouse gas emissions *homo sapiens* will become extinct, perhaps within 100 years. Whatever we do now is too late.' Physicist Stephen Hawking believes humanity can only survive if we leave this planet and colonise another one. There is even a 'Voluntary Human Extinction Movement', whose web site contends that 'human beings are a disease, a cancer of this planet'. They don't advocate mass murder or suicide, but 'non-reproduction, so human beings will eventually disappear' — the world, they consider, 'will then return to a balanced state and evolve more peacefully and naturally.'

Is this 'doom-saying' credible? Maybe, but this book contends there's a sporting chance of exactly the opposite, a large jump forward in almost all respects – new ideas, new inventions of real importance, even some flashes of pure genius, are cropping up all over the world. Many of the practical solutions to our problems already exist, but they languish because it is natural to resist change, and because governments and business are turning their backs on them. Yet we must make the effort to adapt — most of us have children, many have grandchildren; surely we want to hand on a world in reasonable shape to them?

A lot of people worry about the future, but they're not quite sure why. They may know about some of the hazards, perhaps the nuclear threat, energy problems or climate change, but they don't see the whole picture. This isn't surprising — how could everyone find time for the research necessary, consider often contradictory and self-interested opinions, to establish that often unpopular and elusive entity – the truth? Research for this book has taken more than two years, and shown, among other things, that even 'experts' working in specialized areas often see only bits of the big picture.

Next, ask yourself: do I want to know? Surely it's enough keeping up with my job, my home, my kids, friends, getting some fun out of life — nothing wrong with any of that, it's natural for contented people not to like change, we'd rather the world went on pretty much as it always has done. And it's natural to avoid what we don't understand, and what we feel we can't control, so the truly huge issues go into the too hard drawer. 'There are problems, all right, I know, but what can I do about them?' If I've heard this once from people, I've heard it a dozen times. Many people react with this kind of resigned helplessness, while others seem satisfied with simplistic solutions, because they're easier to understand. One of the more common is the *Hunger Games* scenario, in which people, mostly young people, fight to eliminate an evil leader, after which everyone lives happily ever after. This ignores the fact that evil leaders rarely operate in isolation. There are systems behind them, and these need to be dealt with too.

Another simplistic belief is that the answer to the world's energy problems is more nuclear power, even though the facts, including those huge, unsolved issues at Fukushima, still with the potential to destroy millions of people, make it clear this is problematic. Granted much better technology, perhaps using thorium, and adequate safeguards, nuclear may in time be a contributor, but it is impossible for it to completely replace the world's other, highly-polluting energy sources, even in the medium to long term — only if and when fusion reactors are developed successfully could this situation change. These assertions will be expanded and tested in chapter 8.

If you offer facts to people who for one reason or another don't want to know, even facts that are plainly threatening, they don't hear you. This is dangerous, because, in the end, we must deal with those facts ourselves. If governments feel most people don't care about the big issues, they'll continue to do nothing. I can't stress this point too much. *If enough people, the majority, fail to understand and act, we are indeed in trouble* – this is why I rate this human factor as so important. There are bad things out there all right, but the final arbiters, the controllers, have to be us.

Sea level, for instance — there is no doubt that

the oceans will rise, they have done so many times in the past, driven by climate factors similar to those we are grappling with now. A rise of as much as six feet is possible this century, with 70 feet or more in the longer term. Over 150 million people around the world are crowded on to land barely three feet above maximum high tide levels. Much of this low land is made up of islands, but more importantly, it includes the great deltas, the Nile, the Yangzi, the Mekong and many more. Because of their high fertility, these have denser populations than almost anywhere on the planet, and they are where much of the world's food is grown. Long before these flat regions are submerged, storm surges will drive salt water far inland, saturating the land so it can no longer be cropped – this is already happening in Burma and Egypt. Most of these threatened regions are bordered by places already densely populated, so the displaced people will have no way out. They must either die or be resettled elsewhere.

Letting this situation simply develop would have messy and serious consequences. People are not going to sit still and drown or starve, nor should common humanity even contemplate this. But if we do nothing to avoid such a disaster, the large movement of refugees the world is now encountering — there were more than 50 million of them in 2014 — would be nothing to the floods of hungry, desperate people ultimately displaced by sea level rise. And this would not be a sudden one-off catastrophe, but an ongoing, increasing calamity stretching further into the future than we can estimate.

Better, surely, to devise a world plan to resettle these people peacefully and usefully in places that are relatively empty and undeveloped. There are such places, and new technologies are making productive what we've previously considered desert land. After all, we pretty much know already where and when the need for resettlement will arise. The displaced would need to learn the language, history and culture of prospective host nations, and accept training in skills needed by and useful to the receiving country, which would meanwhile build new work and habitat areas. I can already hear the cries of protest — *this would cost money!* Well, so be it. If we want to keep the Steamroller at a safe distance our preoccupation with pure bookkeeping will have to end. Money, after all, is just a marker for work, ingenuity and courage. New productive, motivated people in a country — refugees — generate wealth. That was what made America great.

Change. Right up front is the huge consideration that our world is fundamentally different from what it has ever been, with several potent factors in play that could indeed destroy us all. Wars, weapons, natural disasters, human greed and folly, can be identified as far back in history as we can trace, but never before have we faced vast, apocalyptic forces like climate change, explosive population increase, and huge nuclear arsenals. For every person in the world when I was a child, there are now four. This increase in the number of people is placing an unprecedented stress on land, water and energy resources, so severe that perhaps a third of humans are already deprived of the necessities of life. Six million of those born into our world die every year before the age of five, usually from preventable causes. There are hundreds of millions of angry young men – angry because they can't find work and see no acceptable future. This is not just in the under-developed world. At the time of writing Spain's youth unemployment rate stood at over 50 per cent, in Greece it was 56 per cent. Don't under-estimate this factor as a cause of trouble. These young people need things to do that provide a sense of purpose, and as the inheritors of the future they should –and could — be given a personal stake in dealing with the problems ahead of us.

On the latest count there were at least 16,000 nuclear weapons in existence. Using just a small fraction of these would kill millions; their major deployment might destroy us all. Because the Nuclear Non-proliferation Treaty has failed, and because more and more nations have nuclear bombs, the threat they pose is increasing, not reducing. This is why I rate nuclear war or nuclear accident as the most dangerous hazard of the Steamroller, and accordingly it comes up early, in Chapter 3.

And whatever your opinions on climate change, the facts remain. Sea levels will rise, extreme weather events, floods, droughts, storms, wildfires, will increase, probably to an unprecedented extent. The amount of fresh water in the world is finite, and always has been. In the past it has been ample for the needs of most humans, but now it has to be shared among so many more people there are severe shortages in many places. Climate change, altering where and when rain falls, is compounding this problem. The thin, fragile skin of topsoil – dirt – that provides our food is disappearing quickly because of population pressures. Several nations, including China, the largest, are fighting huge walls of sand blowing in from the encroaching deserts. And everywhere the natural balance between the living things other than us – plants and animals – is being seriously disturbed. Plant life is being sacrificed to the farming of more and more animals, killed for food. And even the resources in the sea – the world's fisheries – are already severely run down.

All bad things, you might say — but there are some good ones, too. There is the Internet, with its social networking offshoots and its pervasive character – consider how successful the Wikileaks site has been in revealing the grubby secrets of the great and powerful. Our means for fast mass communication between individuals on an unprecedented scale will be of the utmost importance in the process of control by ordinary people. Google offers everyone access to information about almost everything, not always reliable, but on the whole useful. There are quite a few organizations that are learning fast how to use these global tools.

And in many ways it is getting to be a better world, slowly, painfully, but definitely advancing. Too many children are still dying young, but only half as many as 20 years ago. Deadly tribal conflicts that have afflicted Africa seem to be a bit more under control, and there are the beginnings of global action to get rid of nuclear weapons. Climate change, once considered a matter of opinion, is now accepted as fact by more and more people. Local, sustainable businesses, models for the future, are cropping up all over the world.

Since no one person could possibly know all the answers to the huge problems of the near future, or forecast with total accuracy the shape of events to come, research for this book has meant tapping into the thinking of as many informed people as possible – it depends essentially on the ideas of hundreds of other people, derived from books, journals and the internet. The objective is to synthesise what they are saying into as clear and uncluttered a narrative as possible.

2 Convergence and Chaos

'More than 140 heads of state and governments flew into New York this week for the UN general assembly amid apprehension that international order is unravelling at an accelerating pace, while the world's leaders seem ever less willing or able to deal with the proliferating threats.' This sentence led the front page of the *Guardian Weekly* in September 2014, the UN secretary general, Ban Ki-moon, saying the world was living 'in an era of unprecedented level of crises.' At around the same time the World Trade Organization pointed out a drop in world trade in that year from 4.7 per cent to 3.1 per cent. The International Monetary Fund in its half-yearly survey in 2014 said the failure of countries to recover strongly from the worst recession of the postwar era meant there was a risk of stagnation or persistently weak activity. Reducing its world growth forecast for that year to 3.3 per cent, the fund warned that the world economy might never return to the pace of expansion seen before the financial crisis.

There was more about the nature of these crises, but not much enquiry into *why* they were happening. This chapter will provide evidence that the Steamroller is already influencing the world, its hazards coinciding with what seem to be random outbreaks of crisis. This should raise a warning — will this contagion continue until the underlying hazards are recognized, and properly addressed? There is very little indication that the international community understands this. I first wrote along these lines more than a decade ago in a book called *The 2030 Spike,* which predicted a disastrous convergence in the 2030 decade of the drivers – basically the big issues already raised in this book. Many of its predictions are already evident.

The ugly image daesh, the self-styled 'caliphate' in the Middle East, presents to the world has prompted an international campaign of bombing the already desperately damaged communities of Syria and Iraq from the air. While without doubt repulsive, daesh is a symptom of larger and more basic disruptive influences in that region. It draws world attention with its fanatical extremism and brutality like the highly publicized executions by beheading. *However, it got its opportunity from major social breakdown in its region.* If there is even a grain of truth in this assertion it follows that allowing the degree of chaos and conflict in Syria and Iraq that bred daesh to continue is highly dangerous; and that aerial bombing and other means of war are not going to eradicate it. Rather, by destroying towns, dams, power-houses they are more likely to perpetuate it, and even worse, make room for other extremist groups that are just as nasty — already the Shia militias seem just as brutal as daesh. Simon Jenkins, in a *Guardian* opinion piece in 2014 says: 'We bequeathed a bunch of warrior zealots a nation in a state of anarchy and a vast arsenal to play with… Their victims desperately need our aid, as do the victims of war everywhere, but not our bombs. For the price of a bombed pick-up truck you can feed a refugee camp for a year.'

The powerful drivers in play seem to be ignored because most people don't yet acknowledge them – or don't want to. Although the effects are often local, they can spill across national boundaries — this is already the case in the Middle East and Africa. This tendency to spread needs to be stopped, just as an epidemic of a disease needs to be stopped, and soon. This will need some new thinking, and major changes in the priorities of the world's largest countries, who should try harder to alleviate at least some of the major problems, like population pressures, poverty and poor government, that have given groups like daesh entry room. As later chapters will show in detail, this is possible, and at no great cost either.

Two international authorities did acknowledge the idea of convergence in 2014. The Intergovernmental Panel on Climate Change (IPCC) reports prior to that year had concentrated almost exclusively on the effects of global warming, but in 2014 presented climate change in combination with other critical areas like population growth, poverty and food security. The report mentioned threats like the dramatic drop in crop yields over the next 50 years in a world that must feed perhaps two billion more people, and warned that climate change, combined with poverty and economic shocks, could lead to war and drive people to leave their homes. In the words of the IPCC chairman, Rajendra Pachauri 'The world has to adapt, and the world has to mitigate.'

The second recognition came with the momentous

and tragic announcement by the UN High Commission for Refugees that in 2014 the number of people forced to flee their homes exceeded 50 million for the first time since World War Two. Of these, half were children, sometimes alone, too often falling into the hands of people traffickers, or to describe them less politically correctly, procurers for the world's brothels. There are various estimates of the number of children in the world sex trade, but all agree it exceeds half a million, and that it nearly always starts with rape, most commonly about the age of 14, but sometimes as young as 7.

The refugee commissioner, Antonio Guterres, said this was 'a world where peace is dangerously in deficit ... There is no humanitarian solution, the solution is political, to solve the conflicts that generate these dramatic levels of displacement.' He said the driving factors included climate change, population growth, urbanization, food insecurity and water scarcity –many of which interacted with and enhanced each other... 'It's sometimes difficult to identify the main motivation... you have a number of people who are forced to move by a combination of reasons, which are not always obvious.' The World Food Programme's ability to feed these displaced people is not unlimited – in 2014 funding shortfalls forced the relief organization to cut food rations by 60 per cent in the refugee camps of southern Chad. A declining economy and a severe drought in Lebanon in 2014 raised fears that country may not be able to feed more than a million Syrian refugees, mostly women and children.

The extent to which society has collapsed in Syria has been an increasing cause of concern to the world. Millions have been driven from their homes, 200 thousand killed, around half of these children, and the country's infrastructure has been severely mauled. Many people, and much of the media, seem to see the issue simplistically as a conflict between the country's dictator Bashar al-Assad and those who oppose him. Yet as the years pass it has become plain that something more than regional and sectarian rivalries is involved. A catastrophic four year drought created massive water shortages and rural unemployment, making more than half of Syria's richest agricultural land unproductive and killing eighty per cent of its cattle stock. This was the worst extended drought and series of crop failures in the country's recorded history, creating food insecurity for more than a million people, who fled into the cities. As a result, the first mass protests of the Syrian Spring in Dara'a in 2011 were among farmers and rural poor fleeing the blighted countryside — climate change and population pressures were the sparks that ignited the years of conflict.

And dry conditions have continued. In 2014 the UN Food and Agriculture Organization said war and drought 'were adding pressure to an already dire food security situation in Syria, raising the prospect of further severe reductions of production of wheat and barley, the country's two most important food crops...the 1.97 million tons of wheat forecast for 2014 is some 52 per cent below the average for the 2001-2011 period.'

But perhaps the most brutal killing zone has been in Africa, in the Democratic Republic of Congo, where as many as six million people are said to have died over two decades of almost continuous war. The converging factors here have been overcrowded living conditions, poverty, political instability, disputes over water and the country's rich mineral endowment, and an almost complete lack of adequate medical services. As many as 90 per cent of those millions of deaths have been from disease – malaria, pneumonia and diarrhoea – and starvation, and more than half of the dead were children under five. Chaotic and lawless conditions have given opportunity to a number of militias, who murder, destroy and rape with impunity. An Amnesty report details looting, often accompanied by torture, killing and rape, systematic pillaging of food aid, stealing from medical centres and planned and co-ordinated attacks and robbing of villages. The Congo conflicts are infamous for the forced recruitment of hundreds of child soldiers, who are often drugged before they are sent out to kill. About 80 million people live in the Democratic Republic of Congo, which is rich in copper, cobalt, gold and other minerals and produces about eight per cent of the world's diamonds.

The failure of Egypt's short-lived democracy and the return of that country to a brutal military dictatorship – brutal enough to condemn 529 of its opponents to death at a single court hearing— are more readily understood in the light of the extreme social and economic pressures there. The strains of over-population and poverty are becoming very evident, with significantly high youth unemployment. The economy is in tatters, heavily dependent on aid money from other Arab states. Egypt has become notoriously overcrowded in recent years, with actual shortages of food during a wheat price spike in 2010. Around a third of the children are stunted from malnutrition. Yet the most recent statistics show that population growth for the two years to 2012 was the largest since records began. More than half a million *more* children were born in Egypt in 2012 than in 2010, Magued Osman, a statistician, saying: 'It's unheard of to have such a jump in a two year period.' However, 2013 set another record, with 2.6 million births compared with half a million deaths. According to one commentator, 'The rising population is seen as a social time-bomb that will exhaust depleted resources and add to social frustration.'

Almost all of Egypt's productive land, on the delta of the Nile River, is only a few feet above maximum high tide level. Within a few decades sea level rise will flood much of that land, while high tide surges make much more of it useless because of salt water inundation. The delta accounts for two-thirds of Egypt's agricultural production, and is home to more to more than 40 million people. There were already reports in 2014 of major problems – in the words of one young farmer: 'The land has become sick. The soil is saline, the irrigation water is saline, and we have to use a lot of fertilizers to grow anything on it.' Eminent Egyptian geologist Khaled Ouda said a one foot rise in sea level would inundate

3000 square miles of the Nile delta and turn 1000 square miles more into islands, isolating towns, roads, fields and industry, while a three foot rise would flood nearly a third of the delta, displacing about eight million people. While the government has attempted to slow the advance of the sea with breakwaters and dykes, scientists have warned the seawater can seep in below them. These problems are compounded by shrinkage of the delta because the 120 million tons of silt that used to come with the annual flooding of the Nile stopped with the building of the Aswan High Dam in the 1980s. The plight of Egypt will be a classic case of convergence, one of the first major manifestations of climate change combining with poverty, over-population and extreme pressure on arable land to create an intolerable situation for millions of people.

In Greece things seem to be getting worse rather than better. The national debt, which stood at 120 per cent of GDP at the start of the financial crisis, reached 177 per cent in 2015. Unemployment rose to a record 28 per cent, socially corrosive youth unemployment reduced very slightly, but remained over 56 per cent – figures described as totally horrific. As always under strains like these an extreme right-wing party has emerged, Golden Dawn, which is steadily gaining more support. Its supporters demonstrating outside the parliament in 2014, some with swastika tattoos, gave the Hitler salute and sang the Nazi Horst Wessel song.

Pakistan will be an early victim of convergence – in many respects it already is. Climate change, manifested by severe annual flooding, over–population, a dangerously frail legal system, severe poverty and insurrection are combining to create a society of increasing chaos. These things are nurturing an insurgency that not only seems out of control, but which is exhibiting a degree of brutality almost beyond comprehension. Such is the Pakistani Taliban. Most societies try to nurture and protect children, even during times of conflict, but in the last week of 2014 Taliban gunmen and a suicide bomber killed 132 children and 17 of their teachers in a systematic pattern of mass slaughter at a school for the children of army officers in Peshawar. This was the worst of a series of shooting and burning down of more than 1000 schools in Pakistan by the Taliban, which wants to replace the nation's fragile democracy with an Islamic state.

Rwanda is our sixth case. In 1994, in one of the worst episodes of mass killings the world has seen, the Hutu tribe murdered close on a million of their neighbours in the Tutsi tribe, although they had previously lived peaceably together for generations in spite of their differences. Jared Diamond, in his book *Collapse,* argues that not only ethnic hatred was involved but also over-population, with too many people trying to stay alive by farming portions of land that were too small… 'Severe problems of over-population, environmental impact and climate cannot persist indefinitely: sooner or later they are likely to resolve themselves, whether in the manner of Rwanda, or in some other manner not of our devising, if we don't succeed in solving them by our own actions.'

Rwanda, with over a thousand people per square mile, is one of the most densely populated countries of Africa – its one and a half million in 1930 grew to over seven million at the time of the genocide, creating pressures that were almost certainly a major element in the disaster. By 2014 there were 11million people in this tiny country. Almost half the children suffer from chronic malnutrition, and life expectancy is 55. More than 70,000 refugees, mostly from the Democratic Republic of Congo, are housed in four camps there.

Afghanistan, which now has some of the worst social indicators in the world, could justifiably be included in this group. The hazards there have been associated mostly with the disastrous and costly war, which at the time of writing was still far from over. According to top US health administrator T.G. Thompson 'war and the Taliban have devastated Afghanistan and its medical infrastructure, and the nation's health challenges are most serious for its women and children. Among its many social deficiencies is by far the worst rate of deaths in childbirth in the world — one in 11 women will die during labour. This is half of all deaths of women aged 15 to 49. One in every four children will be dead before the age of 5. Almost half die from diarrhoea and acute respiratory infections like pneumonia. Half of those who survive are malnourished.

Lester Brown, of Washington's Earth Policy Institute, quoted in *National Geographic,* January 2011, believes there is a serious risk that too little food for the

9billion people who are likely to be on the earth by 2050 will cause a global collapse of civilization. Because we are living off natural capital, destroying productive topsoil and running down groundwater faster than it can be replenished, a drop in food production is inevitable, he believes. Brown called on the world community to unite on virtually a war footing to stabilize climate and keep global population below 8billion – he sees effective family planning as perhaps the most important need. A major case can indeed be made for controlling population – it will be a recurring theme in this book – and could best be achieved by a resolute attack on poverty in the emerging nations, and by understanding that educated girls marry later and have fewer children.

Brown's view is interesting and important because both the major problems he identifies could be dealt with by diverting at least some money from current wasteful or unproductive spending to better purposes. The United States is spending $60 billion a year on nuclear weaponry – much of this is to make 'improved' bombs to replace others regarded as obsolete. The bill for upgrading nuclear weapons worldwide is over $100billion. Unspecified other amounts are needed to deploy these, which, granted the 'best' contingency – that they will never be used – will sit in their silos or in submarines until they, too, become obsolete.

Spending on weapons worldwide in 2013 was well over one and a half trillion dollars. The Global Peace Index for 2014 estimated that the world was spending an astronomical $9.8 trillion in that year on containing and dealing with violence. This is very expensive. If at least some of that money could be devoted to mass production of renewable energy generators and electric vehicles, and to poverty alleviation and birth control programmes, we would be doing much to avert the hideous crisis that might otherwise impact on us all. Perhaps it is salutary at this stage to mention former US President Dwight Eisenhower's warning of the huge power and political influence of the 'military-industrial complex.' The manufacture of nuclear and other weapons is among the largest regions of profit for some of the world's biggest corporations – if you want to sell weapons you need wars, or at the very least, the threat of war.

The poorer you are, the more chance you have of getting sick – this is true almost everywhere in the world. Children can get stomach problems or pneumonia anywhere, but they are major killers in the developing world because their victims are weakened by malnutrition. Those who survive mostly go on to get debilitating sicknesses, usually caused by contaminated water. Rich people can call in a doctor, mostly poor people can't — in the poorest countries doctor numbers might be ten per cent or less of those in wealthier ones. This major difference in average health is another hazard, if only because sick people cannot work as hard, think as clearly as fit people. So fatal areas of chaos can overtake them — mass illnesses, epidemics that spread like wildfire in conditions ideally suited to them.

Other chapters will define the cruel consequences of poverty to the people who are poor. This one should

identify the danger of a disease – it could be Ebola –
moving internationally to the rest of the world from the
poor country that initially harboured it. During the 2014
African epidemic health professionals protested that
although the disease had been around since 1976 little
had been done to control it — it was only in 2014 that
serious attempts were made to start developing a vaccine.

One can say the word 'chaos' and it may not mean
much – hence a closer look at its ugly face is necessary.
The terror group daesh has exhibited degrees of brutality
that deny any pretence of civilization. Soldiers who had
surrendered were butchered en masse, hundreds at a
time, sometimes by shooting, often enough by
beheading. Before death they were subjected to gross
humiliation and torture, after death their bodies were
defiled and dismembered. During revenge attacks Syrian
and Iraqi government forces and militias were just as
brutal.

During the killings in Rwanda the same kind of
deviations from minimal human behaviour were also
evident. The killers were told to spare no one, not even
small babies. Most who died were chopped to pieces
with machetes, not by soldiers but by ordinary citizens,
often enough their neighbours. They were urged on to
kill by local radio stations, which demonized Tutsis as
'snakes, cockroaches, animals.' In Cyahinda 5,800 Tutsis
who sought refuge in a church were murdered. At Butare
University Hospital 170 sick or wounded patients were
dragged out and beaten to death or hacked to pieces. As
almost everywhere else where chaos has reigned rape

was almost universal. In 1994 almost every adolescent girl who survived the killing was raped, with an official estimate of 2000 to 5000 unwanted pregnancies as a result.

So it is evident that reversion back from the cohesion and rule of law of a modern state to barbarism is characteristic of this new phase of chaos. Libya, in which struggle between rival tribes has followed the downfall of Gaddafi, is on the edge of chaos. Many such places get to be run by tyrants, who generally don't want the rest of the world to know what is going on. Although things are getting no better in Syria we seem to hear less and less about that country. This is because of the consistent targeting, killing and imprisonment of journalists trying to work there – such a serious risk that by 2014 many of the world's major news media had withdrawn their journalists and cameramen from Syria.

According to the Frontline Club, a London-based organization of freelance journalists, this attack on journalists has been planned and deliberate. The club has a memorial to Lasantha Wickramatunga, a Sri Lankan journalist killed on his way to work by four gunmen. He foretold his own death in an editorial three days before in these words: 'Murder has become the primary tool whereby the state seeks to control the organs of liberty… Countless journalists have been harassed, threatened and killed. It has been my honour to belong to all of those categories, and now especially the last.'

This has not been an easy chapter to write, nor will it have been pleasant to read, but its purpose has been to show that the Steamroller hazards are real, dangerous and complicit in the world's fast-rising tide of chaos. The following chapters define and discuss these hazards as honestly and clearly as I can manage. Their order is my view of their relative dangers. On that basis the growing risk posed by nuclear weapons comes first.

3 The Destroyer of Worlds

'Now I am become death, the destroyer of worlds.' Nuclear scientist Robert Oppenheimer muttered these words from the Indian classic the *Bhagavad-Gita*, as he watched the explosion of the world's first A-bomb in New Mexico on July 16, 1945, his close associate, Ken Bainbridge, remarking: 'Now we're all sons of bitches.'

Those watching saw a brilliant white light, turning orange as the fireball climbed to a height of almost six miles, where the characteristic mushroom cloud formed. The heat at the bombsite fused sand into thousands of fragments of green glass, and the light flash generated was so intense it is said a blind girl 120 miles away saw it.

Several of the scientists who had helped create the bomb, well aware of its apocalyptic potential, petitioned the US government to quarantine the technology, but this was not to be. Less than a month later, on August 6 and 9, nuclear bombs destroyed the Japanese cities of Hiroshima and Nagasaki, killing more than 200,000 people and bringing World War Two to an end.

The terrifying fate of these two cities brought a shocked reaction from around the world. Protesting marchers demanded that any existing nuclear weapons be destroyed and the technology abandoned, but their pleas went unheeded. The United States and the Soviet Union, global antagonists at the time, each poured billions of dollars into the creation of nuclear arsenals capable of destroying all life on earth. Strategic planning developed into bizarre areas, based on 'MAD – mutually assured destruction', war scenarios were envisaged that were evaluated in mega-deaths – each of these the death of a million people.

As the protest movements grew more vocal, demanding some control over these nightmarish possibilities, US President Dwight Eisenhower, an ex-general, addressed the United Nations in 1953 on the theme 'Atoms for Peace.' He proposed a new international agency to organize the sharing of nuclear information and materials for peaceful purposes, the International Atomic Energy Agency. In 1968 the ill-starred Nuclear Non-proliferation Treaty (NPT) was created.

This treaty aimed at preventing the spread of nuclear weapons by limiting them to the five nations that already had them – the US, Russia, China, France and Britain. These powers undertook to reduce and eventually eliminate their nuclear weapons provided all other countries agreed not to get them. But this objective, reasonable enough on the face of it, was never to be realised. The treaty was signed by 187 nations, although, significantly, not by Israel, India, Pakistan and North Korea. While it still exists on paper and some lip service is given to it, the treaty has substantially failed, and its failure is a potentially fatal result of our inability to organize a planetary body of law.

In the event the original five 'nuclear states' have breached the treaty by retaining stockpiles of atomic weapons that are still a lethal threat to us all. This has not gone un-noticed by the rest of the world – now four other states have nuclear weapons, (they are those that failed to sign the treaty) and several more are believed to have plans to make them.

A four week consultation by 161 nations in 2015 failed to make any progress on nuclear disarmament, although delegates noted that the risk of a nuclear 'incident' was greater than it has ever been, with all four of the 'new' nuclear powers in flash-point regions where conflict is only too likely. The signatories to the treaty also met that year, but were unable to make progress on their major objective — to persuade the five original nuclear powers to give up their arsenals. However, significantly, Israel insisted that regional security arrangements must come before any talks on disarmament. US President Obama offered to negotiate with Russia in 2013 on nuclear disarmament, but without success.

Some technical detail at this point may make the following material easier. Uranium, a heavy metal, is the major source of bomb fuel, but it cannot be used in its natural state because it consists mostly of the substantially inert isotope U238. (Isotopes are forms of the same element whose atoms have a different number of neutrons.) Hence uranium must be 'enriched' to be used in a bomb or a power reactor. Enrichment means the very small proportion of the fissionable isotope U235 uranium contains – 0.72 per cent -- has to be increased to about 20 per cent for nuclear power reactors, 80 per cent or more for bomb fuel. Fissionable means the capacity of the neutron of an atom to split readily and sustain a chain reaction, releasing large amounts of heat and energy -- a fragment of U235 the size of a grain of rice can produce as much energy as three tons of coal or 15 barrels of oil. The most usual method of enrichment takes advantage of the slightly different density of these isotopes, spinning them in gaseous form in centrifuges. This is laborious and expensive, demanding arrays of many hundreds of centrifuges to get adequate results.

The second bomb fuel is plutonium, PU239, which exists in nature only in tiny traces, but which is a major by-product of the operation of 'peaceful' nuclear reactors – it is something we make. Quite apart from its explosive capacity, it is one of the most toxic poisons on the earth and the cancer-producing radiation it emits remains dangerous for at least a hundred thousand years. Once again the process of extracting this from spent reactor fuel rods is difficult, expensive and dangerous.

India and Pakistan – an Islamic country – have been at loggerheads since the bitter and bloody strife that stained the sub-continent when they became independent nations in 1947. They have fought several limited wars over possession of the mountain state of Kashmir, which became Indian at the time of partition because it had an Indian maharaja, although most of its people are Moslems.

India had founded a nuclear research institute even

before independence, in 1948. This developed in 1954 into the Indian Atomic Energy Commission, which by 1959 employed 1000 scientists. It was considerably assisted in its work by the United States and Canada, which provided it with a nuclear reactor able to breed plutonium for nuclear weapons – enough for perhaps one to two bombs a year. The first such weapon was tested in 1974, and others have followed. There are various estimates of the size of India's nuclear arsenal in 2015, ranging from 50 to 120 weapons. Eighty would be a reasonable assumption. This is a force to be reckoned with, especially since India has delivery systems using naval ships, aircraft and rockets. Some of the missiles under development are said to have a range of as much as 6000 miles. While India does not have much uranium, she has vast deposits of thorium – thorium reactors are discussed in Chapter 8.

It is fair to say that Pakistan developed its atomic weapons because its old enemy, India, was doing so. Indeed Prime Minister Bhutto said in 1965 that if India got nuclear weapons 'we should have to eat grass and get one, or build one of our own'. Its nuclear weapons establishment was set up in 1972, and from 1976 was under the effective control of the enigmatic Dr Abdul Qadeer Khan, a German-trained expert on gas centrifuge technologies. This had developed weapons yielding up to 12 kilotons by 1998. Pakistan probably has fewer atomic bombs than India – perhaps 70 weapons – but still enough to effectively destroy its larger neighbour. Pakistani nuclear capability increased significantly with the successful test in 2015 of its Shaheen3 guided missile, which is capable of carrying conventional or nuclear weapons. Its claimed range, 1700 miles, would allow it to reach Israel, as well as most of India.

The Indian nuclear arsenal was developed with the active assistance of the United States, and that of Pakistan by China. The position of the Pakistani Dr Khan is also of interest. Labelled by *Time* magazine (Feb 2005) as 'the world's most dangerous nuclear trafficker,' Khan confessed to making nuclear designs and technology available to North Korea and Iran, both regarded by the US as 'rogue states', but later withdrew this confession. Highly regarded by the Pakistani people because 'he gave them the bomb,' Khan has been convicted then pardoned for his offences in 2004.

The two big south Asian nations have among the most extreme poverty, ill health and child mortality rates in the world. Given the urgent need to improve these things, it seems crass in the extreme to spend billions of dollars and a large part of their scientific expertise on the creation of these obscene weapons. The fact that this is happening, and the many indications that the ordinary people in both countries seem to take pride in their possession of nuclear bombs, point to a level of irresponsibility the rest of the world would do well to take heed of.

In view of the tension between India and Pakistan, it is worth trying to assess the consequences should the 100 or more weapons they possess come into use in outright war. The results would be a major calamity, not only for the region, but also for surrounding nations, even much of Europe, depending on how fallout was influenced by the weather and wind direction at the time. A worst case event could result in millions of deaths.

In Israel there was – and probably still is – a subterranean nightclub expressly designed as a cave. With its dim lighting and crags of artificial rock it did not seem to me particularly attractive. However it was popular enough — all the tables were taken on the night I was there. Were all these people attracted to this grotto because it seemed a cellar-like refuge — and from what? This was all supposed to be a happy Saturday night out, a carefree social occasion, but it didn't feel that way. The people here seemed to be a little less than happy. There was a definite air of tension.

When I talked with people later on in that visit it became plain that Israelis see themselves as in a state of siege, surrounded by enemies. This has bred a climate of fear, a certain inability to see the point of view of 'the enemy' – the surrounding Arab nations — and an unexpected amount of aggression. When I talked to a senior diplomat about this over drinks — why were Israelis like this? I was told: 'There are Jews all over the world, very nice people most of them – but the extremists are here.'

Israel refuses to say whether she has any nuclear weapons, but there is no doubt she has, probably more than 100, and the means to deliver them. When I asked a very high-level official specifically about this I was told: 'We don't have such weapons, but we have the means to make them, and we could do that very quickly if we had to.' The same informant added: 'We see nuclear weapons as a last resort. They are our safeguard – ensuring no-one will dare attack us.'

'We shall never again be led like lambs to the slaughter.' This statement, attributed to Ernst David Bergmann, who was the chairman of the Israeli Nuclear Energy Commission from 1952 to 1966, is a clear enough indication of Israel's motivation to build a nuclear arsenal. American author Seymour Hersh described Israel's policy on nuclear weapon use as 'the Samson Option.' It has been summarized, apocryphally, since I can find no clear source for the quote: 'If we must go down in flames then, by God! We shall take this whole damned planet with us!' Attributed to former Israeli Defence Minister Moshe Dayan with more certainty are these words: 'Israel must be like a mad dog, too dangerous to approach.'

These quotes, and others like them, make it clear Israel has nuclear weapons, or at least the bomb components that can quickly be assembled. Her policy of 'nuclear opacity', neither denying nor admitting that her weapons exist, is accordingly difficult to accept. In this context, Israel's attitude to Iran becomes horribly dangerous. Iran fired up its first nuclear reactor, ostensibly for peaceful purposes, in 2011. Many Israelis seem to think Iran intends to use it to build nuclear weapons, and some believe the Israeli air force should strike pre-emptively and destroy the reactor. Iran's response promised all out retaliation in which Israel might be destroyed. This mutual hostility has persisted – – too likely a scenario for the Samson Option for the world to feel safe.

Israel began to explore the Negev Desert for uranium less than a year after the new state was founded. Some, but not much, was found in phosphate deposits. In that same year, 1949, six physics graduates were sent overseas to study nuclear science. The British and French attack on Egypt after that country's President Nasser nationalized the Suez Canal significantly advanced Israel's nuclear opportunities. When she agreed to enter the conflict, advancing across the Sinai Desert, the French and British offered a *quid pro quo* — generous assistance with Israel's nuclear programme through the later nineteen fifties and sixties. France provided a research reactor and some nuclear fuel for the establishment that was set up at the isolated desert site of Dimona in 1956. This EL-102 reactor was said to be capable of producing 22 kilograms of plutonium a year.

When in 1961 the then Prime Minister David Ben-Gurion announced that Israel planned to build a plutonium separation plant it became clear to the world that she intended to make nuclear weapons. That plant was completed in 1965, and the first bomb was probably assembled in 1967. Probably four to five warheads were produced at Dimona every year, considerably more than that as time went on. The 'probables' in this paragraph are because the Dimona plant was and is shrouded in very tight security and secrecy, the government releasing no information about what is happening there and continuing to claim its work is for peaceful purposes only.

Considerable light was shone on this when former Israeli nuclear technician Mordechai Vanunu made revelations that occupied most of the front page of the London *Sunday Times* and was subsequently carried in newspapers around the world. Vanunu, who had worked at Dimona for nine years, confirmed that an active weapons programme was going on there, including secret separation of plutonium.

I met Vanunu briefly when he was living in Sydney, Australia just before his disclosures to the media, and was left in no doubt of his sincerity as an anti-nuclear activist. He was already afraid of the Israeli secret service, the Mossad. I told him he should be safe if he kept out of Israel. In the event I was quite wrong. While in Sydney Vanunu met a *Sunday Times* journalist, Peter Hounam, and travelled to London with him. The newspaper was cautious about Vanunu's revelations, asking nuclear experts to confirm them. They subsequently did. However, the Mossad had been tipped off. In a sequence of events that might have come straight from a spy novel it sent one of its woman agents, Cheryl Bentov, to London to become acquainted with Vanunu. Posing as an American tourist named Cindy she persuaded Vanunu to go with her on holiday to Rome. Once there, in a totally illegal act, the Mossad drugged him and spirited him back to Israel by boat. There he was convicted on treason and espionage charges and gaoled for 18 years, eleven of these in solitary confinement. Having served that sentence he remained in 2015 under virtual house arrest, forbidden to leave Israel, talk to journalists, or to publicly state the anti-nuclear views he has always professed. He has been arrested and imprisoned again several times, the most recent in 2010. Vanunu has been nominated for the Nobel Peace Prize several times, and is considered a prisoner of conscience by Amnesty International.

Vanunu's statements, and the ferocity with which he was pursued and punished, amount to *de facto* confirmation that Israel is a major nuclear power. By researching a consensus of informed opinion around the world it is possible to make a reasonable assessment of what this huge force amounts to. Some commentators put it as high as 400 weapons. While the hundred or so that seem most likely are negligible compared with the thousands held by the United States and Russia, the Israeli arsenal, if fully used, would represent a grave danger to the world. Because the Middle East is so central to the Eurasian landmass and because of the continuing violent confrontation with Palestine and Iran, hundreds of millions of people on that landmass are at perpetual risk.

Israel's most advanced ballistic missile, the Jericho111, was said to have become operational in 2008, with MIRV capability. This means the missile can carry six 100 kiloton missiles, which can be independently targeted to different locations, or a single one megaton weapon. Its range may be as much as four thousand miles. She also has five – possibly six – modern German-built Dolphin-class submarines, which are believed to carry nuclear-capable cruise missiles, and specialised 'black' air units designed to deliver nuclear weapons. All this means that Israel has the power to hit any target in the Middle East, Europe, and Africa, with a reach possibly extending into Russia, parts of India, China, and western Canada and the United States. Even if the homeland itself were destroyed, submarines at sea would be able to launch 'vengeance' second strikes. This seems integral to the Samson Option.

There has been a good deal of public noise about the possibility that Iran intends to build nuclear weapons. This may well be true, but there is no evidence yet she actually has them, or is even close to having them. Similarly, although it has generated huge amounts of publicity, North Korea's nuclear capability seems much smaller than any of those described above, and accordingly ranks lower on any risk index. After announcing in 2002 that it proposed to build nuclear weapons, North Korea commenced a programme of plutonium separation, testing one nuclear device in 2006 and another in 2009. The Russians rated these at about the size of the Hiroshima bomb – perhaps 15 kilotons, while American experts estimated the yields at just a few kilotons. North Korea is thought to have produced enough plutonium for five to eight weapons in that range. Accordingly, the threat from these weapons is regional, causing concern to its neighbours, South Korea and Japan, rather than to the world at large.

However, the threat from the other three 'small' nuclear arsenals is worldwide. Just what would happen if there were a 'limited' war between India and Pakistan, or Israel and its Middle East neighbours, which involved the use of 100 Hiroshima size weapons? Would the rest of us just look on? The Internet is crowded with speculation on this theme. The consensus is:

*Millions of immediate deaths in the warring nations in areas at and near ground zero, and millions more in their neighbours from radiation fallout.

*A huge smoke and dust-cloud rising initially six miles high, then still further into the stratosphere, where absence of rain could ensure the sun remained obscured for several years. This cloud would cover so much of the world it would cause an immediate drop in global temperature, which some believe could be as much as seven to eight degrees Celsius. However, even a three to four degree drop would be enough to return us to conditions in the Little Ice Age from the 14th to the 19th century, when millions starved to death. Since world population is so much greater now than then, the fact that virtually no food could be grown for several years would prompt famine on a huge scale, with perhaps hundreds of millions more deaths. This famine would be exacerbated by the contamination by radiation of large areas of farmland.

*Massive depletion of the ozone layer, greatly increasing world incidence of skin cancer and cataracts.

*Huge dislocation of world transport and trade.

*The loss of much of the world's art and culture, destroyed in the ruined cities.

This damage bill should be enough to give all of us pause. We must somehow neutralize and then eliminate the nuclear weapons in India, Pakistan and Israel. This would take many years and could only be achieved by an unprecedented degree of public agitation, followed by global co-operation offering these nations absolute guarantees as to their security. This would be difficult, but not impossible, in spite of some formidable block points. For instance, India has built her nuclear arsenal not only because she fears Pakistan but also because she fears China even more.

While the world would be severely mauled by one of these 'limited' wars, it and the human race could survive, although in considerably smaller numbers.

Of course the weapons described above are only a tiny fraction of the 16,400 in the world. The rest are held by the five 'big' powers, the ones who promised to eliminate their nuclear arsenals and have not done so. There are a lot fewer than ten years ago – only a third as many, but the difference is largely academic, since there are still more than enough to destroy the world several times over.

The world's thousands of nuclear weapons have spanned a wide range, from the now discontinued low yield Davy Crockett, an American field weapon light enough to be hefted by a man, to the world's largest nuclear explosion of the Russian Tsar Bomba, with a yield of 50 megatons, 50 *million* tons of TNT equivalent. War which involved the use of the current range of weapons would kill billions of people quite quickly, then most of the rest of us around the globe as the world's wind patterns moved the deadly fallout – the scenario depicted in Neville Shute's novel *On the Beach*. That book concluded that everyone in the world would die. This may not quite be the case. There might be just a few survivors, who would have to live for many years, possibly even many generations, in underground bunkers where they had exclusive access to massive stores of preserved food. They might number a few thousand, at the outside.

What to do? Mass demonstrations, civil disobedience have failed to solve this one — most of us can only shout from the sidelines and hope that the leaders of the nuclear powers will somehow come to understand how dangerous this situation is, and act decisively to rid the world of the monster we have created. This would only be possible if absolute guarantees of their territorial integrity could be given to the smaller nuclear powers —global nuclear disarmament seems unlikely without this.

There have been quite a lot of close calls with nuclear weapons. By far the most dangerous was in 1961, when a crippled B52 bomber jettisoned a hydrogen bomb 260 times as powerful as the Hiroshima weapon on the US state of North Carolina. The bomb's four arming systems activated except for one faulty switch.

That small, defective device was all that averted a nuclear explosion that would have killed hundreds of thousands in the eastern United States. The full facts on this incident were only revealed in 2013, under freedom of information.

In 1962 a man called Vasili Arkhipov was second officer in a Soviet submarine off Cuba, which was being attacked with 'practice' depth charges. The captain of the submarine, believing the attack was for real, ordered a nuclear torpedo launch against the US aircraft carrier *Randolph*. The political officer, one of three required to agree, supported the captain. Arkhipov dissented — his action probably averted all-out nuclear war. At three in the morning on November 9, 1980, a faulty computer chip triggered a nuclear attack warning on the US. The fault was discovered before retaliatory action planned by the White House could begin. There were several similar incidents in 1979 and 1980, Defense Secretary Harold Brown commenting that false warnings were virtually inevitable.

These, and a dozen or more like them, suggest that if a nuclear exchange begins, mechanical failure or human error are more likely to be the cause than a deliberate act of war. A prime danger of an accidental nuclear strike beyond its immediate effect is the machinery of automatic retaliation. This 'fail deadly' mechanism, which all the nuclear powers are believed to have, provides for a massive return strike even if command and control structures in the attacked nation have been destroyed.

Given global agreement to the abolition of nuclear weapons it would take many decades to dispose of huge amounts of radioactive material. Nobody seems to know just how much of this there is, except that it amounts to many thousands of tons, mostly from dismantled 'obsolete' weapons. However the American Institute for Science and International Security estimated about 1855 tons of plutonium in 35 countries in 2004 — enough to make 255,000 nuclear weapons. Since a lot of weapons have been dismantled since then and nuclear powerhouses create another 30 tons or so of plutonium every year, a likely 2015 figure would be well over 2000 tons.

Many of the world's biggest repositories are so badly managed they are dangerous. Fred Pearce, writing in *New Scientist* in 2015 said 'decaying structures are cracking, leaking waste into the soil and are at risk of explosions from gases caused by corrosion' at Britain's Sellafield Reprocessing Centre. Researchers are still uncertain what to do with 90,000 tons of radioactive graphite stored there. Cleaning up the site, due to end in 2020, is costing almost two billion pounds a year.

The Doomsday Clock is a device used by the *Bulletin of Atomic Scientists* to define the risk to the world of nuclear warfare and climate change. Nineteen Nobel Laureates contribute to its estimate. Midnight on the clock indicates the end of civilization. Early in 2010 it was moved back one minute to six minutes before midnight, but was returned to three minutes to twelve in

2015 because of the rapid advance of climate change. According to the bulletin 'World leaders have failed to act with the speed or on the scale to protect citizens from potential catastrophe. These failures of political leadership endanger every person on earth.'

4 Getting Warmer

Queen Victoria's consort Prince Albert was in the chair at a meeting of the Royal Institution in London in 1859 when the existence of greenhouse gases in the air and their effect on warming was demonstrated. Polymath Professor of Natural Philosophy John Tyndall showed that while heat passed almost unhindered through the basic gases of the air, others blocked heat almost as effectively as a panel of wood. These, carbon dioxide, methane and water vapour are now, 150 years later, recognized as major greenhouse gases, preventing infra-red radiation from the sun escaping back into space, much as happens inside a greenhouse. CO_2 is measured in parts per million of the atmosphere – ppm — currently standing at a little over 400. This contrasts with pre-industrial levels of 280ppm.

A first take on what global warming means might best be to consider how the climate has been lately. In 2012 a staggering 93 per cent of more than 900 natural catastrophes in the world were weather related. They cost the world $170 billion. A single event, Hurricane Sandy, which devastated New York, came with a bill of more than a billion dollars. According to Oxfam the number of weather-related disasters tripled over 20 years to 2013, during which United Nations figures estimated $2 trillion in economic loss, with 1.3million deaths, 4.4billion people affected. Those statistics represent an appalling volume of human suffering and frightened, homeless, hungry people.

And what of the future? A carefully considered statement by the three organizations representing most of Britain's climate scientists (the Met Office, Royal Society and Natural Environment Research Council) late in 2009 warned that 'long-term changes in climate will persist for millennia.' This is because the oceans, which drive climate, once heated, give up that warmth only very slowly.

Twenty fourteen was the hottest year on record, and 13 of the 14 warmest have occurred since the turn of the century – not dramatically so, but resulting in average global temperatures measured on land, in the sea and in the air of one degree Centigrade higher than they were 100 years ago. Disquietingly, ocean temperature rose 0.66C. Why? While most climate scientists say this is because we've been putting more greenhouse gases into the air, others believe global warming is part of a natural cycle. They instance an earlier warm period around the 10th century — although recent research suggests this was regional, rather than global — and a gradual recovery from the Little Ice Age since about 1830, as evidence for this. Both these factors – the natural cycle and greenhouse gases – are probably involved. What has to concern us is the *extent* to which human influences are responsible. After all, we can do something about those. The natural cycle is beyond our control.

Whatever the reasons, the over-riding reality is

that the world is warming more rapidly than natural trends could account for, and that the extreme weather events of recent years are becoming dangerous. So just how much has global average temperature risen? According to the US National Climatic Data Center, in January 2014 land temperature was 1.17degrees Centigrade above the 20th century average. This figure is, of course, an average. In regions in or near the Arctic, like northern Canada and Alaska, the rise was above 3degrees, indicating very rapid climate change. The rate of increase seems to be growing – warming of .13 degrees a decade in the past 50 years, and .18 since 1976. However, during the decade to 2014 the warming influence was almost flat – around 0.07 degrees C.

At first blush these figures might seem small, and the apparent recent check in global warming reassuring. However, there is a general scientific opinion that a two degree increase would be dangerous, and four degrees or more catastrophic. We have already warmed the planet more than one of those two degrees.

By 2015 the picture had changed dramatically, with a considerable body of informed opinion predicting a rise of 4C. Following a conference at Melbourne University in 2011 the Associate Professor of Environmental Policy there, Peter Christoff, edited a symposium of its findings, published in 2013. Speaking on a radio programme Christoff said: 'Four degrees unfortunately is a very realistic prospect by the end of this century –perhaps as early as 2070.'

Meanwhile journalist Mark Lynas had put together

a book visualising a four degree or more increase, with huge sea level rises, up to 50 per cent less water available in South Africa and the Mediterranean area, a planet completely free of ice for the first time in 40 million years, disastrous reduction in food supplies, much of southern Europe becoming a desert. In 2012 the World Bank, certainly not an organization dominated by wild-eyed greenies, published a lengthy and carefully researched report 'Turn Down the Heat; Why a 4 degree World Must be Avoided', predicting ' a world of almost unimaginable social, economic and ecological catastrophes,' and saying 'we hope that this report shocks us into action – the scenarios are devastating; the inundation of coastal cities; increasing risks for food production leading to higher malnutrition; many dry areas becoming drier – wet regions wetter; unprecedented heat waves; exacerbated water shortages; more frequent, more intense cyclones; irreversible loss of biodiversity, including coral reefs.'

This report pointed out that an average world temperature rise of 4C meant somewhat less than that over the oceans, but higher over land. This could range from 6 to 10 degrees in the sub-tropics. Since those averages include the night, day temperatures could peak even higher, so high they could rise beyond human endurance. We lose heat by sweating, but there is a limit to this. Humidity, the amount of moisture in the air, is critical to how much heat a human can tolerate. A temperature of 36C with humidity at 70 per cent can be just as dangerous as 42C in drier air at around 15 per cent. If inhabited areas now experiencing 45 degrees at times are exposed to 6-10C more, they could plainly no longer support any reasonable human presence. In January 2014 two of Australia's largest cities, Melbourne and Adelaide, had unprecedented heat waves, with five days continuously over 41C and peaks of 45. Analysts reported 167 'excess deaths' in Melbourne as a result. There were huge firestorms, so violent experienced fire officers at a subsequent enquiry considered them beyond control by any means. Neither city is tropical; they both lie well within the temperate zone.

Pakistan's nightmare summer of 2015 ought to be a warning to the world. Morgues and hospitals overflowed as thousands died from heat strike when temperatures rose into the mid forties with high humidity. Ten thousand more were being treated in hospital. Power failures affecting air conditioners, fans and water supply, together with the unusual heat, contributed to disastrous conditions, especially in Karachi, a city of 20 million people. The city came virtually to a standstill as the Sindh state government declared a state of emergency and several days of public holiday so outdoor workers could stay inside. A heatwave in India a month earlier killed 2000 people.

Abnormal weather is already coming close to destroying whole nations. Super typhoon Haiyan, the strongest ever to make landfall, killed more than 6000 people in 2013, mostly in the Philippines. It destroyed more than a million homes, displacing more than 4 million people. At least 15,000 survivors were still living in tents when a second major storm, Typhoon Hagupit, struck the same region barely a year later, forcing almost a million people into emergency shelters again.

Huge hurricanes in 2008 swept across the Caribbean island nation of Haiti, wiping out almost all of the food crops that normally feed nine million people. In 2013, five years after the monstrous typhoon Nargis hit the fertile Irrawaddy delta in Burma, half a million people were still without adequate housing, with tens of thousands without access to fresh water and totally dependent on international food aid. The storm, one of the worst experienced anywhere, killed 140 thousand people and destroyed 800,000 houses. A 12foot high storm surge, driven by 135mph winds, sent a flood of salt water through huge areas of rice paddies, destroying their fertility.

A catastrophe in Tabasco Province of Mexico began in the Atlantic as an incipient hurricane, but degenerated into a tropical depression that caused days of unremitting torrential rain. Floods extended over 80 per cent of the province, driving a million people from their homes. A few months earlier 200,000 people lost their homes and their fields when Hurricane Felix swept through Nicaragua. Huge winds and a 15foot storm surge destroyed rice paddies and vegetable plots in Bangladesh, killing more than 3000.

Torrential rain in Pakistan in 2010 killed 2000 people and displaced 14 million, as well as destroying huge areas of the country's most productive cropland.

There have been heavier than usual and destructive monsoon rains there every year to 2014. The Pakistan government has already estimated it will take years for the country to recover. Meanwhile there is the terrifying prospect that the flood rains of three successive years might be due to a permanent distortion to the pattern of the monsoons. This goes beyond mere speculation – over the last decade these seasonal rains have indeed been heavier. There has also been some well-informed opinion on this. Lord Julian Hunt, former director-general of the Meteorological Office in London, says 'The danger is that Pakistan, and the Indian sub-continent in general, will become the focus of much more regular catastrophic flooding… this trend is fuelled by global warming, and potentially by any intensification and alteration of the el Nino-la Nina cycle.'

These events and opinions surely offer a clear warning – and they were just some of the wild storms and extreme weather events that are now becoming painfully regular, and which are forecast to worsen as world temperatures rise. One of the more important things to remember about these disasters is that while they disappear from the news cycle in a matter of days, recovery takes many years, during which time thousands of people suffer profoundly.

Greenland, the world's largest island, is regarded by scientists as something of a canary in a mine, an indicator of climate change trends. Settled and farmed by the Vikings a thousand years ago, it was abandoned in the early 15th century as snow and ice closed in with the onset of the Little Ice Age — a cold period that lasted until the 19th century. Now, with warmer weather its formerly tiny population is growing, cattle and sheep are raised, vegetables grown, and huge catches of fish taken.

However, this recent human activity happens close to the coast. Go inland and you will soon come across the Greenland Ice Cap, the largest deposit of land ice in the world after Antarctica's. Around seven per cent of global fresh water is locked up there. This lens-shaped ice cap covers 80 per cent of Greenland, and in places is more than two miles thick – its weight is so great it has depressed the central part of the island into a basin thousands of feet below sea level. Like the Antarctic ice mass, Greenland's is constantly on the move, spreading out towards the coasts under its own weight, and spilling more fresh water into the sea through huge glaciers, which are speeding up and 'calving' icebergs into the ocean. Since this is land ice, sea levels globally would rise more than 20 feet if it all melted. Fortunately this is not likely to happen any time soon, although eventually it will, if global warming continues.

While floating ice does not raise sea levels as it melts, it does affect global warming indirectly. As the white ice is replaced by dark-coloured sea the albedo – reflectivity – increases. The darker colour absorbs more of the sun's heat, raising the local sea temperature and the overall melt rate. Ice reflects 85 per cent of the sun's heat back into space, open sea only 7 per cent. This is why the unprecedented rate of Arctic ice melt in 2007 was so important, leaving as it did the smallest area of ice since satellite measurements began in 1979 — 1.6 million square miles — and opening the North West Passage to conventional shipping for the first time. Melting in subsequent years was slightly less severe, with the ice at 1.7 million square miles at the end of the northern summer of 2015. Much of this ice is, however, getting thinner, so the actual volume is reducing faster than the area.

Around a quarter of the land in the northern hemisphere – almost 9 million square miles – has permafrost. This is soil that stays permanently frozen, although it is normal for the top surface to melt in summer and refreeze in winter. Now, as it melts more in summer and takes longer to refreeze, greater volumes of fresh water are entering the Arctic Ocean, and much more carbon dioxide and methane are being released into the atmosphere. Methane —'marsh gas' — is formed as organic matter decomposes in boggy soils.

Climatologist David Lawrence, from the US National Center for Atmospheric Research, has concluded that the big permafrost regions in Alaska, Canada and Russia could triple their rate of warming during periods of rapid sea ice loss. This would involve a very large increase in methane from the hydrates locked in the ice. Since methane has a much greater warming effect than carbon dioxide, this would speed up climate change. Organic matter dating back to the era of the dinosaurs is now exposed, ready to surrender its huge masses of methane. The researchers also reported: 'This situation, when summer thaw extends more deeply than the previous winter's freeze, can lead to a *talik,* which is a layer of permanently unfrozen soil sandwiched between the seasonally frozen layer above and the permanently frozen layer below. A *talik* allows heat to build more quickly in the soil, hastening the long-term thaw of permafrost.' Melting beyond averages is already happening. Among its consequences is major damage to infrastructure, roads, bridges and building foundations, which are collapsing in almost all permafrost areas. Sections of the recently built high-altitude railway from China into Tibet are already threatened.

The East Siberian Ice Shelf in the Russian Arctic is estimated to contain 50million tons of methane. In 2011 Russian scientists who had been studying methane 'plumes' in this region reported that release of the gas had increased alarmingly, in the words of the team leader, Dr Igor Semiletov: 'In a very small area, less than 10,000 square miles, we have counted more than 1000 fountains, or torch-like structures, bubbling through the water column and injecting directly into the atmosphere. We carried out checks at about 115 stationary points and discovered methane fields of a fantastic scale – I think on a scale not seen before. Some of the plumes were a kilometre or more wide – the concentration was a hundred times higher than normal.'

This research and other findings like it have prompted estimates of the cost to the world of future methane releases. Researchers from Cambridge University in England and Erasmus University in the

Netherlands call it 'an economic time-bomb.' Under business as usual conditions, with greenhouse emissions at present rates, the researchers estimated the cost of damage caused by methane from East Siberia alone at $60*trillion* dollars. This figure, representing more than three quarters of the entire global economy for a year, seems huge — however, it has been put forward by responsible and informed researchers.

By 2014 the dimensions of this problem had been confirmed by several other studies, including a finding from the Russian team that the seafloor off the coast of Northern Siberia is releasing twice the amount of methane than had been previously estimated. Research also showed that much more methane hydrate exists in vulnerable shallow Arctic waters than had been thought previously, and that a global temperature rise of 1.5 degrees — half a degree above present readings — could trigger massive releases. A Canadian study of 71 wetlands around the world found many permafrost regions were already transforming into wetlands with high methane emissions.

Then there is an emerging threat that dangerous viruses that have been sleeping for thousands of years in the permafrost might be released among humans who have no resistance to them. In 2014 French researchers brought such a dormant virus back to life. *Pithovirus* came from permafrost 32,000 years old. The largest-sized virus yet known, it is still virulent, destroying amoeba it infected in 12 to 15 hours. It does not seem to attack humans, but the Marseille scientist leading the team that discovered the virus, Chantal Abergel, said 'there is good reason to think there could be pathogenic viruses out there too.' The more permafrost melts, the greater the risk, which would also be increased if mining operations disturbed the soil.

In the words of eminent economist and climatologist Nicholas Stern, in a *Guardian* article in 2014: 'Delay is dangerous. Inaction could be justified only if we could have great confidence that the risks posed by climate change are small. But that is not what 200 years of climate science is telling us. The risks are huge.'

So what to do? We face an unprecedented level of crises if global warming becomes extreme, so we must see that never happens. Every suitable property should have solar water heating and power generating panels, and houses should be planned to be energy neutral. Fossil-fuelled ships, cars, aircraft should give place to non-pollutant electric vehicles. Coal-fired power stations must be phased out. Farming animals for food must be severely reduced, if not eliminated totally. Use of food crops to make ethanol must stop. I know this single paragraph is asking for pretty much a transformation of society, so is it justified?

'Severe, widespread and irreversible impacts for people and ecosystems' — these are the words the International Panel on Climate Change's fifth report used to describe the likely outcome if we continue to emit greenhouse gases. This report issued late in 2014, says:

'Heat waves will occur more often and last longer, extreme rainfall will become more intense and frequent, the ocean will continue to warm and acidify, and global mean sea level will rise.'

People have heard much of this before, but the real kick in this document, buried well down in the small print is this: 'Without additional mitigation efforts beyond those in place today... *warming is more likely than not to exceed four degrees above pre-industrial levels by 2100.'* I tried to read or listen to as much expert comment on this report as possible, and was especially struck by the reply an eminent lady professor made to a question about what we'd do if warming went over four degrees. She simply said: 'We won't be here.' That's it, in the starkest possible terms, for as much as half the human race.

There has been an understandable reluctance to visualize just would happen, but it is something we should face squarely. Even the most optimistic opinions foresee an almost total breakdown of civilizations and the collapse of almost all societies. Others warn of the extermination of many populations from starvation and disease. The amount of arable land would steadily reduce as whole regions turned into searing desert, disappeared under the sea, or were poisoned by salt-water seepage. Food production would be reduced to probably under half present levels. Developing countries would be hit hardest as more people starved, the ancient killing diseases would soar out of control. Some researchers are predicting world population could drop to one billion, while almost all say it would return to medieval levels, perhaps four billion. Accurate estimates are impossible, because the world has never faced this situation before.

Here are some extracts from a variety of informed opinions: 'Ocean ecosystems and food chains would collapse… half the world would be uninhabitable… likely population capacity: under one billion people… we are on the precipice of a great tipping point, an area of no return… are we talking about how we might adapt to a four degrees warmer world? Have we gone mad?'

There is little doubt that any talk about adaptation to four degrees or more is dangerous nonsense. That minority who survived such a catastrophe would do so in a very unpleasant and hostile world, where they struggled for the bare necessities of life.

But all this need not happen. We have to come to our senses and do whatever it takes to stop greenhouse emissions rising, even if this means we have to go to bed earlier, travel less, walk more, live in the country, turn vegetarian, grow most of our own food, get our hands dirty. Scientists have done their homework, and we know that to have any hope of keeping temperature rises under two degrees we can put no more than 600 gigatons of carbon into the air by 2050. Five times that amount would be emitted if we were to use all our known resources of hydrocarbons —that would result in catastrophic temperature increases. Emissions were 44 gigatons in 2014 — a 2.5 per cent increase on the previous year. This does not mean we can afford to pollute at current levels for another decade, because there are major emissions we can't directly control, like the unpredictable amount of methane release from permafrost melting. And the hundreds of new coal-fired power stations scheduled over the next several decades in India and China will increase the greenhouse gas load, not reduce it.

In practical terms, we have no more than a decade to stop using the hydrocarbons. Four to five years would be a prudent target.

5 Denial, and the Evidence for Climate Change

Since all aspects of the climate change debate have become politicized, there is a need to look at the evidence for global warming. This is all the more necessary because some climate-denying 'think tanks' that claim to be authoritative have a vested interest in blurring the issues, at best. Many of these have been funded by the big fossil fuel miners. This statement needs to be backed up by reliable evidence, and this comes in some detail in a 2007 paper *Smoke, Mirrors and Hot Air* compiled by the Union of Concerned Scientists – 'ExxonMobil, the world's largest publicly traded corporation, doesn't want you to know the facts about global warming. The company vehemently opposes any governmental regulation that would require significantly expanded investments in clean energy technologies or reductions in global warming emissions…The corporation has spent millions of dollars to deceive the public about global warming' by funding a complex of 'fronts' – apparently legitimate organizations which in fact all serve a quite small number of 'sceptical' scientists, most of whom have no specialist expertise in climate matters.

But ExxonMobil is by no means alone. Energy multinationals selling oil and coal all over the world have contributed millions more to this same cause of disinformation. One of the largest campaigns, in 2009, used intensive lobbying of the United States Congress to sabotage climate change legislation brought forward by President Obama. Former vice president and US senator Al Gore should know something about the way American politics work. In his 2013 book *The Future* he says: 'The carbon fuel companies hired four anti-climate lobbyists for every single member of the US Senate and House of Representatives in their fight to defeat climate legislation. They have become one of the largest sources of campaign contributions to candidates of both parties. Many of these companies have provided large sums of money over the last two decades to "liars for hire", who turn out a seemingly endless stream of misleading, peripheral, irrelevant, false and unscientific claims.'

Why? Even the briefest of studies answers this query. ExxonMobil, for instance, is huge — its $40 billion a year profit exceeds the gross domestic product of most of the world's nations. Every week it can continue to exploit the hydrocarbons is worth a small fortune, so it is in its financial interest to discredit climate science. It is, however, responsible for vast amounts of pollution, according to the Union of Concerned Scientists – 'the end use of its products in 2005 resulted in 1047 million metric tons of carbon dioxide-equivalent emissions. If it was a country, ExxonMobil would rank sixth in emissions.'

So here is the evidence for climate change, falling broadly into two areas – scientific research of past climate events as indicated by ancient fossils and ice cores, and observation of what is happening now. Was the inhabited world ever dangerously, even catastrophically hot? The answer is yes, so hot that for millions of years both poles were free of ice and animals and plants now more typical of the temperate zone flourished there. When Ernest Shackleton and his pioneering expedition struggled towards the South Pole in 1908 they found seams of coal and the fossilised remains of wood and tree leaves. For some time continental drift was thought to have explained this evidence of past vegetation – once Antarctica was located far from the poles and had a temperate climate. But that was a very long time ago. Research now suggests Antarctica was 'green' as recently as 40 million years ago. The continent had already been at or near its present position for 50 million years then.

The alternative explanation could be that the climate was much warmer – and evidence coming forward is tending to support that. An extraordinary surge of global warming 55 million years ago is estimated to have raised temperatures five to eight degrees. Sea level was at least 60 feet higher, crocodile-like creatures lived in an Arctic Ocean that was apparently a warm freshwater lake in which the water temperature was as high as 18C. There is evidence that high levels of greenhouse gases were responsible, due either to volcanic eruptions or massive releases of methane or both, but there is no reliable way of confirming events so long ago. However, earth's climate did not return to normal for 200,000 years. Some climate scientists take this as a warning of what might happen again if the level of greenhouse gases continues to rise.

More recent events, up to a million years ago, are more readily established, because much of their history is locked up in the great ice caps. While only a little under ten per cent of the earth's surface is permanent ice, it amounts to almost 87 per cent of all fresh water, and in places is many miles deep. Ice 'coring' carried out over the last 30 years involves deep drilling to produce long cylinders of ice which, when studied, can establish carbon dioxide levels, temperature and sea levels over hundreds of thousands of years.

Deep, enduring ice in the polar and some alpine regions is the result of snowfalls over innumerable years. While this snow is mainly frozen water, it contains traces of other elements in dust, volcanic residues and gases trapped in air bubbles. These last can identify the nature of the atmosphere, and through their position in the core, the time when the snow fell. Even the temperature at that time can be deduced by examining the molecular nature of the air. A mile long core taken in Antarctica in 2004 contains snowfalls going back 740,000 years, encompassing eight ice ages and their intervening inter-glacials. Another taken at the Russian base in Antarctica dates back almost half a million years. While this seems straightforward enough, in practice it is not. Simply getting the ice cores in cold and hostile regions is difficult, arduous and dangerous, working in icy winds and temperatures as low as 40C below zero. The 'contaminants' in the ice occur only in tiny amounts, and it requires much knowledge and great care to identify what they mean.

However, as the evidence from drilling starts to accumulate, it has shown a clear association between higher rates of greenhouse gases in the air, higher sea levels, and an elevated average temperature of the earth. This does not necessarily confirm cause and effect, nor do these factors always correspond in time, there is no unequivocal evidence that elevated greenhouse gas levels have always prompted global warming. The most that can be said is that these three factors tend to be associated.

Some climatologists point out that volcanic eruptions and variations in the heat output of the sun and the tilt of the earth must have been, and still are, contributors. But whatever the influences, during the last major interglacial between 116 and 130 thousand years ago, when global temperatures were at least three degrees higher than they are now, sea level rose almost 20 feet. This could have assisted our beginnings — very early fossil remains of *homo sapiens* found in caves near the mouth of the Klasies River in South Africa date back 118,000 years. It can be surmised that these primitive humans, who used crude stone tools and fire and gathered plants and seafood, prospered in the warmer climate before moving out of Africa into the Middle East, Asia and Australia, and eventually to Europe.

Marine fossils of shells and coral, even ancient animal teeth can be and are analysed to establish climate fluctuations. Annual bands in coral can be read just as annular tree rings are. The evidence from these things confirms the huge swings in earth's climate, and the rise and fall of the sea over the ages.

These studies have established a definite climate pattern – over the last half million years ice ages have lasted around 90,000 years, while warmer inter-glacials, like the one we are now in, have typically persisted about 10,000 years. Since our interglacial is already beyond that, some climatologists say we can expect another ice age some time over the next few thousand years. It will come – it is only a matter of when – unless extreme global warming defers it.

The second major indicator of global warming is present-day events on the ground, such as the size of the polar ice-sheets and the state of glaciers. These huge rivers of ice are normally in a state of balance, losing melt-water to the seas at about the same rate as replacement by snowfalls. This 'normal' state has, however, been disturbed during the modern industrial era. Many glaciers around the world have been retreating, their faces moving further up their valleys because of warmer weather. Most of the world's important rivers begin at the foot of glaciers, and are dependent on their stability for a regular supply of water. These great rivers of ice act as regulators, catching and storing snowfalls, then releasing water dependably. If they disappear this situation of balance will change into a succession of extreme floods and droughts. The big rivers of Asia, like the Yangzi, the Indus and the

Mekong, depend for their water on the thousands of glaciers in the Himalayas. Many of these glaciers are in retreat, sending larger amounts of water downstream for the time being. However, if this continues river flow will be greatly diminished. Tens of millions of people who rely on this water to produce essential food would be in danger of starvation, perhaps within decades.

The other major indicator is reduction in the amount of ice at the poles. In terms of geological time this ice has until recently remained stable, but this is no longer the case. In 2002 a huge mass broke away from the Larsen B ice shelf on the eastern side of the Antarctic Peninsula because unusually warm weather had allowed temperatures to rise above freezing. The most recent events in this region – and they are startling enough – are described in Chapter 13.

The situation in the Arctic is quite different – even dramatically so, with some climatologists predicting that the north polar region could be completely ice free in summer by 2030. This is suggested by a much more rapid loss of ice in 2007 than had been expected. Following extensive satellite observations the European Space Agency reported the summer extent of the ice had shrunk by more than half a million square miles, almost ten times the average over the previous decade. Previously unknown islands appeared, and shipping companies speculated about using the Northwest Passage – open for the first time in many thousands of years – to reduce journeys between Europe and Asia by more than 1000 miles. The six nations bordering the region also joined into an unseemly jostle to exploit oil and gas fields now unlocked, and to gain territory – Russia even claimed the North Pole and Canada the North West Passage.

There is a form of ice you can hold on the palm of your hand and set fire to. It burns readily with a clear flame. This is the 'new' hydrocarbon, methane hydrate, an arcane substance with the consistency of sorbet, resulting from water and natural gas combining under high pressures and low temperatures. There is an awful lot of it under the polar ice caps and in deep water on the edge of the continental shelves. Even conservative estimates put the amount at 200 thousand trillion cubic feet, perhaps five times present natural gas reserves.

Methane hydrate could be a major player in the climate debate because methane is such a potent greenhouse gas. The methane, highly concentrated inside lattice-like cages of water molecules, expands 160 times when liberated. In normal climatic conditions it stays put, but if you warm things up, it is likely to get away. This is one of the dreaded 'tipping points' one hears about. If enough methane escapes, it could create a positive feedback situation where more and more is released, creating a situation beyond our control. Methanes from the arctic permafrost are now being released at an increasing rate — this is one to take quite seriously.

Would you be happy with a world in which it was

never again possible to look up and see blue sky? This could be a consequence of one of the high tech, impractical and highly expensive ideas now being touted as 'geo-engineering'. The theory behind these is that if, as it seems, we can't get our act together enough to control global warming, there might be some 'magic pill' scientific solutions that would save us the trouble. Almost all are aimed at reducing the amount of sunlight. One proposes launching a million square miles of mirrors into space – it is said this would reduce solar radiation by one per cent. Another takes its cue from volcanic eruptions, during which millions of tons of sulphur can be ejected into the atmosphere. The proposal is to do just that. These ideas have been criticized because of their huge expense, the risk of increasing the amount of acid rain and possible damage to the ozone layer, to say nothing of the grey skies worldwide that could result. According to Reading University researchers they might also cause a colossal drought in the tropics.

Somewhat more feasible but still immensely costly is the use of seawater sprays to increase cloud cover. Another proposal, again needing big money, is to freeze carbon dioxide, enclose it in giant torpedos, and drop it to the bottom of the sea. These 'technofixes' have been criticized as just another smokescreen, designed to delay effective action to reduce emissions.

The world has put together an expert group of scientists, the Inter-governmental Panel on Climate

Change, to keep us informed on what is happening, and what we need to do. Founded in 1988, the panel consists of 1500 authors and expert reviewers, although 2500 scientists from 130 countries are involved in one way or another. Its last major report was in 2014.

The panel has been heavily criticized by creatures of the large multi-national fossil fuel energy companies in the US Congress and elsewhere, who repeatedly seek to discredit climate change science by whatever means. The 2007 report was lengthy and detailed, and it was perhaps inevitable that some errors would occur.

However, none of these affect the major findings, which are broadly the facts and trends outlined in this chapter.

What then, were the errors in the 2007 report? It claimed that the Himalayan glaciers would disappear by 2035. This was acknowledged to be wrong – the figure should have been 2350. Some minor errors of a statistical nature were detected in the so-called 'hockey-stick' graph depicting past temperature history.

Although attempts were made to discredit the entire report, and, indeed, the IPCC itself, an enquiry made at the request of Congress by the prestigious US National Academy of Sciences confirmed it as substantially correct. Indeed, other critics complain that the report was too conservative, and that climate change was now more advanced than it stated.

The world is indebted to the patient and thorough work of the thousands of scientists who make up the IPCC, and would do well to heed what they say rather than be confused and misinformed by a malicious and manifestly partisan barrage of criticism. In the introduction I assessed the attitudes of individual people as the main risk to the human species – those responsible for that campaign of misinformation are very near the top of my list. This is not to say that climate science should not be challenged – it should, but by genuine sceptics who know the science rather than the deniers.

So what does the IPCC have to say? Here are some findings:

*The major greenhouse gases, carbon dioxide, methane and nitrous oxide in the air have increased markedly since 1750 and now far exceed pre-industrial values.

*The amount of carbon dioxide in the atmosphere in 2014 — 399 parts per million (403.21 in May 2015) exceeds by far the natural range for the past 650,000 years, which was 180-300ppm.

*The amount of methane in the atmosphere, 1774 parts per billion, exceeds by far the natural range of the past 650,000 years, which was 320-790ppb.

*Cold days, cold nights and frosts have become less frequent. Hot days, hot nights, and heat waves have become more frequent.

*Observations since 1961 show the ocean has been absorbing more than 80 per cent of the heat added to the climate system, and that ocean temperatures have increased to a depth of 9800 feet.

*Average Arctic temperatures have increased at almost twice the global average rate over the past 100 years.

*Mountain glaciers and snow cover have declined on average in both hemispheres.

*There has been an increase in hurricane intensity in the North Atlantic since the 1970s, and this correlates with increases in sea surface temperatures.

A recital of these factors is like a tolling bell – surely it would take someone very obtuse not to understand that if even some of them develop we are in for bad trouble – bad enough to avoid at almost any cost.

6 The Coming Great Flood

The *Wall Street Journal* (13.7.09) compared the G8 leaders' resolve at their summit in Italy in 2009 that global temperature must not rise beyond 2C with King Canute's alleged command to the tide not to rise. Actually the 11th century monarch has had a bad press – – it was his courtiers who said he was so powerful he could even forbid the waters to rise. Canute sat on the beach and got his robes wet to teach them a lesson – 'Let all men know how empty and worthless is the power of kings.' Seven years after that G8 resolve, with the average temperature of the world likely to go up by 3C, even 4C, the oceans will rise everywhere – two feet, three, even six feet? Nobody knows, for sure. However in 2013 the World Meteorological Organization reported that the rate of rise had doubled from that of the 20th century to three millimetres a year, with the seas on average about eight inches higher than in 1880. This has been caused by ice melting at the poles and thermal expansion of warming seawater.

A 2015 report from the Potsdam Institute for Climate Impact, while agreeing with the current annual three millimetre figure, says sea level rise was more rapid over the last two decades than had been previously thought. This reassessment, resulting from over-estimates of the rate of rise during the 20th century, indicates that rather than doubling, it was probably closer to two and a half times that in the 20th century.

The vice-chairman of the IPCC, Professor Jean-Pascal van Ypersele, remarked at a policy conference on sea-level and ice sheets in 2010 that recent satellite observations were 'starting to show, but quite convincingly, I must say, that both the Greenland ice sheet and the Antarctic ice sheet are losing net mass, not only on the margins, but on the ice sheets as a whole.' Subsequently the IPCC 2014 report predicted almost three feet of sea level rise this century, with West Antarctica a wild card, likely to push levels well above that figure, to a possible five feet.

In 1900 San Marco Square in Venice flooded six times a year; now it happens as much as 50 times, with residents and visitors becoming accustomed to *acqua alta* –high water – hotels make rubber boots available to visitors, street vendors sell plastic leggings that fit over your shoes. While most of the flooding is due to subsidence – the entire city is sinking slowly – sea level rise is also a significant contributor.

After the tiny Pacific island nation of Kiribati Bangladesh is perhaps the country most vulnerable to sea level rise and extreme flooding – the Dhaka environmental group Coastal Watch says 11 families there are losing their homes to water erosion every hour. There are forecasts that 40 per cent of the productive land in the southern regions would be lost from a two foot rise by 2080, but much has been affected already by sea water seeping in, turning, in the words of one farmer 'into a huge saline swamp where no vegetation grows.'

Coastal land in Egypt and several south-east Asian states is also being poisoned by salt to the extent where it is no longer productive. In many places seawater is invading aquifers, the location of most of the world's fresh water – – in some Pacific island states, almost all of it. A major evacuation plan is proceeding to get the people out of the low-lying Carteret Islands of Papua New Guinea, already visibly affected.

These are just a few of the reports of damage caused already in many parts of the world by this manifestation of climate change. But in spite of these early warnings the world is doing too little to prepare for one of the more dramatic and dangerous events of the current century. At best this intrusion of the oceans will cause the gradual destruction of beachfront property and flooding of low-lying land on islands and in river deltas – this best-case scenario is bad enough, involving as it does millions of people. At worst, on responsible evaluations, the rising sea could deprive as many as 150 million people of their land and means of life within 80 years. Almost all are in regions that are overcrowded and resource poor, ill equipped to deal with this flood of climate refugees. And this would be just a beginning. The steady rise of the oceans will go on, because of the sea's enormous capacity to retain heat, for hundreds of years, as it has done many times before.

Eventually half a billion people could be displaced, and many of the world's major cities could suffer from major and prolonged flooding, including its most cherished, like Venice and St Petersburg. The speed and severity of this new Great Flood will be in direct proportion to our ability and resolve to deal with greenhouse gas emissions quickly and effectively. The more we argue and delay, the less we do, the faster and more disastrous the consequences will be.

The rate at which the great polar ice-sheets melt is critical to this gigantic process. If they were all to go the seas would rise an appalling 240 feet, almost all of the world's major cities would be inundated, and the planet's ability to feed a greatly reduced population would be severely restricted. Most of the ice is around the South Pole – the majority of those 240 feet would come from there. If the Greenland ice fully melted this would raise the sea more than 20 feet. Another extreme gives some indication of the vastness of the sea's potential movement – at the height of the last ice age it was 400 feet lower than it is now.

These total melts are unlikely inside time frames in the hundreds of years, even if global warming becomes acute, although there could be 'tipping points' beyond which we would not have the power to control melting. In that case rises in the hundreds of feet would indeed be likely eventually – inevitable even.

Sea-level rise varies in different places, the highest being the low latitudes and the least at the poles. This is attributed to the effect of winds, powerful ocean currents, the gravitational pull of the polar ice sheets, and the subsidence or the 'rebound' from the burden of ice of some landmasses. (There is also the geoid effect, which would take too much space to explain here – look it up if you're interested.) As the volume of ice at the poles decreases, so will be the amount of water their gravity pulls towards them — that water will increase sea level elsewhere. There are definite 'hot-spots' of higher rise like the north-east coast of the United States — the New Jersey shore can expect a three foot rise — and much of Europe. Professor Aslak Grinsted, of the University of Copenhagen's Centre for Ice and Climate, who has studied the dynamics of sea level rise in the North Sea, Baltic and north Atlantic, is predicting rises of as much as five feet for England and the Netherlands this century.

The world average rise of about an eighth of an inch a year – a foot in a century – hardly seems intimidating, but less reassuring is the growing evidence that this is more than twice the average for the 20th century. Will this increase continue, and how far? The rate of ice melt is being studied with great care to establish just this, because the rise and fall of the sea is so directly associated with the amount of ice in the world. During ice ages more water is locked into the expanding icepacks, and the level of the sea falls. During inter-glacials the warmer sea melts much of this ice and sea levels rise. There is no doubt about this; the archaeological record clearly shows that over the ages there have been regular massive fluctuations in sea level. Warmth, average world temperature, is the key — the temperature of the oceans is especially significant. This is why global warming and sea-level rises are associated.

However, since we know that even in the most extreme climate conditions that could be foreseen the great ice packs would take a long time to melt, why worry? There are two reasons – the first is that even slight to moderate melting would cause sea-level rises severely disruptive of world society because so much productive and highly populated land is low-lying. The second is that some parts of the ice-masses are melting faster than the rest, with solid evidence that this may soon increase alarmingly. This is the case in the Arctic and in the West Antarctic Peninsula. These melt-water rivers abruptly drop into almost vertical sinks called *moulins,* which take the water right through on to the underlying rock shelf, where there is a theory it might act as a lubricant. This could cause the whole West Antarctic shelf to move faster down the slope.

The evidence is hardening for average rises of at least three feet this century, with two feet much sooner. Anders Carlson, of the University of Wisconsin-Madison, and colleagues have studied the rate of ice loss from the ice-sheet that extended as far south as New York during the 2000 years after the end of the last ice age. A study of beryllium isotopes in ancient bedrock led them conclude that the sea rose between two and four feet a century during those millennia. Climatic conditions then were very similar to those we can expect for the rest of this century. If these researchers are right this spells bad news for the millions of people who live on land that is less than three feet above maximum high tide level. According to the IPCC's 2007 report a rise of that extent would 'affect' at least 250 million in the Asia-Pacific area alone. Dr Atiq Rahman, a Bangladeshi lead author for the IPCC, has predicted that 35 million people in coastal areas of his country would be dislocated by 2050. Britain has given over $100 million to a special fund to help Bangladesh cope with climate change, and several other wealthy countries are expected to contribute.

These are rather cold facts and figures. I invite you to consider a relatively small island, Bhola, lying between two channels of the Ganges River where it runs into the Bay of Bengal. Bhola, around 100 miles by 20, is home to 1.8million people. Most of it is barely three feet above high tide level. As the glaciers in the river headwaters melt faster more water rushes down, cutting into the fringes of the island. There are houses and gardens right up to the riverbanks, and these are disappearing regularly into the muddy waters. An unusually high tide in 2013 flooded much of the island, displacing half a million people and causing crop losses valued at $10 million. The IPCC predicts the rising sea will destroy 17 per cent of Bangladesh by 2050, displacing 30 million people, but Bhola is likely to have disappeared totally by then.

The low-lying Ganges River delta extends into the neighbouring Bengali lands of India. Here 50 million more people are at risk. The most vulnerable are the four million who live on the Sundarban Islands, marshy flats threaded with a tangle of tidal creeks and rivers. Fifty-eight of the 108 islands are inhabited. The rest make up a reserve of swamp forest and waterways people enter only with caution – they are one of the last retreats of the

Bengal tiger.

Low-lying coastal land like this is among the world's most productive – the huge river deltas, like those of the Nile, the Ganges, the Mekong and the Yangzi have always been areas of high population, and are very fertile, producing around half of the world's food. Delta countries like these would be severely affected by a three foot rise and flooded almost completely by six feet.

The Netherlands Delta Commission, established to define the risk of climate change, said in its 2008 report 'this is not a nightmare scenario, but a reasoned view of the future.' Predicting sea-level rises as high as three feet this century and six feet within 200 years, it has recommended spending more than $20 billion over the next 20 years to strengthen dykes and levees by a factor of ten, and to provide especial protection to the giant port of Rotterdam. The money will be spent mainly on bolstering existing seawalls, but also into pumping vast quantities of sand to build up dunes, and digging minor waterways and canals to channel flooding. Dutch engineers estimate they can control the situation by these means up to a sea level rise of five feet. After that much more radical solutions would have to be considered, such as altering the course of the Rhine River and building artificial shelter islands off the coast to reduce storm surges. One of these surges in 1953 killed almost 2000 people in Holland.

By late 2008 the Dutch government had

announced that it anticipated costs of $1.6 billion a year for the next 100 years to combat climate change. This was based on estimates of sea-level rise as high as four feet this century, at least twice that in the next. By 2010 an extraordinary range of projects, from complexes of floating houses and greenhouses to research into salt-tolerant agriculture, had started.

Because they were originally shipping ports – most still are – some of the world's biggest and most important cities will be at risk quite early. Over half China's people, 70 per cent of its large cities, nearly 60 per cent of the national economy, are located along her coastline. London, New York, Bangkok, Dhaka, Tokyo, Mumbai, Ho Chi Minh City, Lagos, Vancouver and beachside suburbs in scores of other cities will be affected.

Most of the world is now populated more heavily than it has ever been, so the need to resettle perhaps half a billion people from flooded areas could trigger a crisis of unprecedented proportions. Hence while it seems that sea-level rise is as yet so slight most of us needn't worry much about it, this impression is illusory – we are pushing levers now that cannot be returned to their safe positions. While the majority of this generation of adults will suffer little of consequence over the next few decades, our children later in their lives, and especially our grandchildren, will not be so fortunate. This is because once it warms, the ocean will not cool again for hundreds of years. Even if we were able to cut back our greenhouse emissions now, this would still be the case.

One of the world's best-known climate scientists, Dr James Hansen, who heads NASA's Institute for Space Studies, said 'I find it almost inconceivable that "business as usual" (induced) climate change will not result in sea level rises in metres within a century – it's a much more dangerous problem than we realised… if we get warming of two or three degrees Celsius then I would expect both West Antarctica and Greenland would end up in the ocean.' Dr T. Scambos, a glaciologist at the US National Snow and Ice Center, warned that Arctic ice is melting much faster than the IPCC predicted. More than 50 scientists from four countries involved in the Andrill Project, ice drilling near the South Pole, reported that carbon dioxide concentrations only slightly higher than they are now would result in a significant loss of permanent Antarctic ice. This conclusion is based on a 1280 metre long core taken from the Ross Ice Shelf.

This brings us to the concept of the 'tipping point.' Think about an old-fashioned set of beam scales — a weight can be pushed some distance along the beam and nothing happens. Then if it is pushed just a tiny amount more the beam will suddenly tip – there will be a very large movement. Hence once the sea warms to a certain extent its rise will just go on — we won't be able to stop it. Many scientists now believe this will also happen with the polar icecaps — once melting reaches a certain point it will continue, regardless of our efforts to check it. This is because the combined load of albedo, methane emissions and rising average temperatures becomes so great it overloads the system.

While we are stuck with sea level rises for hundreds of years, they will happen slowly, so we have time to act. If we don't act effectively the present significant number of refugees will grow into a torrent. It isn't difficult to visualise hundreds, maybe thousands, of boatloads of climate refugees, hungry, poor and desperate, trying to get a foothold in those countries they believe have space for them. These people probably would not speak the language of their target country, know little about its culture or way of life, and have no skills of value to their reluctant hosts. Public opinion in the target countries would without doubt be hostile, regarding the newcomers as an unwanted liability. But the newcomers, flooded out of their countries by carbon emissions they had done nothing to cause, would insist they could scarcely be expected to drown or starve to death.

Mutual hostility could grow into conflict quite quickly. Naval forces might be sent out to turn unseaworthy boats around, exposing the people in them, including women and children, to severe risks. Those who did get a foothold might be desperate enough to fight to stay, using any weapons that came to hand. And while these conflicts raged, the world's economy would be severely damaged, and its food supply fatally reduced. World trade would fall off, piracy would undoubtedly flourish and pitched battles would be fought all over the planet for possession of any land that could produce food.

Is this what we want? If anything, the foregoing

paragraphs probably understate the chaos and violence that would result if we just let the situation drift.

But it doesn't have to be this way. We are an enlightened, compassionate and intelligent species, aren't we? There has to be another way, and of course there is.

Just as we have set up an international body to control commerce, the World Trade Organization, the first necessary step could be the foundation of a World Sea-rise Authority, with wide-ranging powers, adequate finance, and complete freedom to act. The sooner this is done, the better. There will be problems within a matter of years from now.

The authority's primary responsibility would be to organize the resettlement of the millions of the dispossessed in an orderly and constructive way. This could begin with a major world public relations campaign, using all of the media, and in particular, schools, to inform people, especially in potential host countries, about the problem and its inevitability. Next would come methodical surveying to establish who are likely to be affected first. These people would then become early targets for an information programme to prepare them to be acceptable and useful migrants.

Granted they accepted this discipline, world enquiries could begin to establish which countries might be prepared to take them, how many, and on what terms. Education of the prospective migrants in the language and culture of their future hosts, and training in skills sought and required by the host country, would be mandatory. Financial and technical assistance would be needed by the host countries to set up industrial and agricultural enterprises, housing and other social infrastructure so the migrants could become useful workers and citizens as soon as they arrived. A planned increase in agriculture on under-used land – especially new and innovative projects – could absorb many of these people, and do much to alleviate the grave food shortages that sea level rise will cause.

Controlled and adequate transport would be needed to move the migrants at appropriate times. Suitable accommodation would need to be provided for the disabled and ill, orphans and the elderly. In co-operation with national governments, there must be worldwide restriction on development of land plainly at risk, without exceptions, other than wharves and similar facilities for sea travel. All this would be associated with careful monitoring of the rates of sea-level rise to identify vulnerable areas and the times at which movement of people becomes necessary.

Surveying and engineering programmes would be designed to convert unused land in the flooded countries to productive agriculture – this could include engineering of steep land, perhaps by terracing, improvement of poor and depleted soils, and more intensive use of existing farmland. Programmes to reclaim land in the inundated regions could begin. Sixty-five per cent of the Netherlands is below sea level. What can be done there can be done elsewhere, given time — but not everywhere. Most of the affected areas would be poor, and unable to afford expensive seawalls and pumping stations.

This raises the very large, bottom line question – even for those who can afford them, are protective barriers the best option? Already billions of dollars are being spent or committed to projects designed to barricade regions or cities against the rising water. Four big Asian cities with a total population of more than 50 million are already planning to build barricades and seawalls to keep back the rising water, which is resulting from a combination of subsidence and sea level rise. In 2007 a strong monsoonal storm coinciding with a high tide over-ran the flimsy coastal defences of the Indonesian capital, Djakarta. The resulting storm surge covered half the city with water as deep as 13 feet and caused damage estimated at more than half a billion dollars. In 2013 this disaster was repeated, leading to inception of the Great Garuda Plan to build a 15 mile outer seawall and 17 artificial islands at a cost of $40 billion. The project, described by the government as 'one of the most challenging hydraulic civil works to be carried out in the world', is due for completion in 2022.

One of the more desperate cases is Ho Chi Minh City (formerly Saigon) in Vietnam, which is almost completely flat, built on land that is mostly 18 inches to three feet above sea level. Plans for a dyke and floodgate system were announced in 2008, but little seemed to have been done when the city experienced major flooding in 2012 and 2013. Dhaka, in Bangladesh, had 2.8 million people in 1981 — in 2015 there were 13 million, many living in low-lying areas already flooded by high tides. Bangkok, in Thailand, is another low-lying delta city. Record flooding there in 2011 not only inundated the capital but impacted 65 of the nation's provinces as the water surged inland. With damage estimated by the World Bank at almost $50billion this was considered one of the five costliest natural disasters in modern history.

And it's not just Asia that's facing trouble. A US Geological Survey report in 2012 predicted that raised sea level along the American east coast will be four to five times the global average in this century, with rises of as much as six feet, making New York, Boston and Norfolk vulnerable. Miami is already in trouble because it is built on porous limestone, which is allowing seawater to seep through the city's sewerage system.

Many billions of dollars have already been spent or are committed to floodwater defences. This may solve immediate problems, but in the medium or long term, might not the money be better spent on resettlement of displaced people? A number of people are thinking about what will happen in the 22nd century. Almost all believe the oceans will go on rising, as much as 30 feet in the long term, and that many of the world's cities will have to be abandoned. Some speculate that wealthier people may retreat to artificial islands - effectively huge ships each able to accommodate 100 thousand people and designed to stay at sea permanently. This might work for a minority, but would be of scant use to the bulk of humanity.

Sentiment favours trying to defend our cities, but plain common sense says otherwise — spending billions on physical barriers is a costly mistake.

The rise in sea level may not be the only dangerous influence of global warming of the oceans. The most dramatic predictions have involved changes to the flow rates of ocean currents, which are one of the major engines of climate. Climatologists are concerned that the Atlantic Conveyor, which includes the Gulf Stream, might slow down. These currents keep temperatures in northern Europe and northern America liveable by bringing warm water north. As this warm, salty water gives up its heat it normally drops to the ocean bottom and is returned southward at that lower level. Because this massive ocean system is influenced by the amount of fresh water flowing into the Arctic seas, the rate of ice melting in the north polar region again becomes significant.

While there has been a certain amount of panic publicity depicting a Britain and northern Europe locked in ice with disastrous crop losses, stopping or even hesitation of the ocean currents remains speculative. There is no clear evidence of any permanent slowing, and the flow rates fluctuate naturally anyway. But if there were even a small effect it might add to the cold winter tendency already apparent in Europe and North America for other reasons associated with climate change.

In 2014 researchers at the University of Exeter returned this issue to attention, saying: 'We found that natural fluctuations in the circulation were getting longer-lived... the continued influx of fresh water, driven by global warming and the melting polar icecaps, could be enough to slow it to a halt.' Probably a greater risk is that changes in the ocean currents could disrupt the Asian monsoon, which brings annual rain more than two billion people depend on to grow their food. This cloud-laden wind provides Asia at large with more than half its rainfall, India with 90 per cent. These torrential downpours normally come after months of drought, and there have been severe famines when they fail or weaken. And a more intense monsoon can be equally disastrous, as the record recent flooding in Pakistan and throughout South-east Asia has shown.

A huge flying saucer seems to be grounded on the small, beautiful, but now uninhabited island of Runit in the Pacific's Marshall group. Actually this flattish concrete dome, 350 feet across, is a repository for large amounts of radioactive waste resulting from the scores of American atomic bomb tests in this region in the 1950s. The island is off limits for people more than 24,000 years, since much of this debris is plutonium, which will remain dangerous for at least that time. Originally designed to be temporary, the Runit dome is cracking, but the Americans don't want to cope with it and the Marshall islanders can't afford to. However, these islands are only a few feet above the ocean, and are already affected by sea level rise. A little more sea level rise and a single large typhoon could at some time fracture the dome, and its contents would spill into the Pacific.

The Runit dome is just one part of a large problem caused by the proximity of so many nuclear power stations to the sea. The World Nuclear Association says 'the simplest method (for cooling) is to run a large amount of water though the condensers in a single pass, and discharge it back into the sea.' This is why between 50 and 100 nuclear plants were, like Fukushima, built close to the sea – 12 of 19 in Britain are at risk of flooding from sea level rise.

7 The Energy Dilemma

Scarcer and more expensive energy is an increasing concern, but is it one of our more serious problems? After all there are scads of coal seam gas everywhere, and while crude oil's getting harder to extract, there still seems to be enough of it. Sensible cars run on half as much fuel as they used to, so what was all that about peak oil? We seem to be getting more frugal in our use of energy—according to the UN Environment Programme the advanced industrialized countries used almost one per cent less electricity in 2014 than the previous year.

However, the issue isn't so much about immediate availability as the kind of fuel we burn. Our massive dependence on hydrocarbon fuels, especially coal, creates the carbon dioxide that triggers climate change. The world's two largest countries, China and India, have a huge appetite for this black and dirty mineral to build the infrastructure and provide the electric power they need to raise the living standards of their people. Between them they account for more than half the global demand for coal, which is cheap compared with other sources of energy. However, because it's almost pure carbon, burning coal, especially brown coal (lignite), creates this troublesome load of carbon dioxide.

This worries a lot of people. Climatologist James Hansen, in an open letter to US President Obama in 2008, warned that the continued use of coal would set carbon dioxide levels so high as to guarantee destruction of much of the life on the planet… in his words 'coal plants are factories of death.' So we are between a rock and a hard place. The faster we burn the hydrocarbons – and that goes for oil and gas too — the quicker carbon dioxide in the air goes up, the more rapid the increase in global warming. China brings a new coal-fired power station on line every two weeks, overtaking the United States in 2007 to become the world's largest emitter of CO_2. India is now in third position – more than 70 per cent of her electricity is generated using coal, and she plans to burn much more. Even worse, three quarters of the world's coal fired generators are 'dirty', with low temperature burning that results in almost twice as much CO_2 as the more modern 'supercritical' plants that use powdered coal and high temperatures.

Over the last 30 years billions of dollars have gone into 'clean coal' research, achieving little other than to confirm there is no such thing as clean coal at a competitive cost. Much of this research has been into geosequestration, which means collecting the carbon dioxide and piping it long distances to caves or abandoned mines. This has proved to be prohibitively expensive and could even be dangerous. Carbon dioxide escaping from a fractured pipeline into low-lying country could pool, killing everyone in that region. That has happened before. In 1986 Lake Nyos, in the Cameroons, emitted a large cloud of carbon dioxide of volcanic origins that killed 1700 people, most of whom died in their sleep.

Conventional oil, which powers most of the world's transport, air, land and sea, is much closer to running out. An Eos report from the American Geophysical Union in 2013 said flows from existing fields were falling five per cent a year – this confirms the result of several other studies – and that production from 'unconventional sources' like shale was difficult and expensive, with a very low investment return. In 2015 falling oil prices and increasing protest at the very polluting nature of Canadian 'tar' sand mining had forced the closure or suspension of at least 40 projects — one think tank, Carbon Tracker, saying an oil price of $95 a barrel would be needed to make any new mines economic. It is also increasingly unlikely that subsidies to Canadian oil, $3.5 billion so far, will be maintained.

A five per cent a year fall in production means a drop of half in conventional oil in a single decade. And there is another disquieting possibility. If there were war or some sudden civil crisis in the Middle East, where most of the world's remaining oil is located, there would be an almost immediate and serious reduction of the fuel and food supply in many places.

Because of these risks changing the ways we produce energy should be well up on our list of priorities. Ideally we need energy forms that won't eventually dry up, and we need to start building them soon, in large quantities. Developing alternatives takes time and money – as least half a trillion dollars to replace the energy provided by fossil fuels, with lead times of 20 years at best.

So we'll need transitional fuels that produce less greenhouse gas. They do exist, but can't be relied on to keep us going indefinitely. Some existing coal-fired power stations are already being converted to burn natural gas, which is still plentiful, and generates about half the CO_2 that coal does. However this does not take into account the amount of 'fugitive' gas that escapes during the production process, which some researchers claim makes gas almost as pollutant as coal.

Compressed natural gas can power ships and vehicles – if we want to be sure food continues to be in city shops we should get on with converting our delivery fleets as fast as possible. Fuel ethanol is already being used for motor vehicles in many countries, but should we be making it from sugar cane and corn, which are food crops, in a world getting short of food? However, ways of making ethanol from plant residues that would otherwise be waste are in the pipeline, as also is research into making it in algae ponds. There are cost and land use problems with both of these, so they could scarcely replace oil as a fuel completely.

Electricity produced by falling water – hydropower – is one of the most important sources of future power. It is essentially non-pollutant but does have some disadvantages. Between them the Asian nations are building or planning more than 400 dams on the great rivers of that continent. The largest of these, the Three Gorges Project on the upper reaches of the Yangzi River in China, has an installed capacity of 22,500 megawatts, larger than that from 20 nuclear power stations, and is the basis for a huge expansion of industry and cities in south western China. Costing around a third of a trillion dollars it features the world's largest ship-lift, a series of locks than can raise ships as big as 14,000 tons from the river below the dam to the lake above it. The environmental group International Rivers construction says this involved the flooding of 13 cities, 140 towns and 1350 villages and displaced more than 1.2 million people – the reservoir behind the dam is a virtual inland sea 400 miles long. "The environmental impacts of the project are profound… a festering bog of effluent, silt, industrial pollutants and rubbish, erosion, landslides downstream, threats to one of world's biggest fisheries in the East China Sea and a probable contribution to recent droughts.' International Rivers warns that now China has gained the knowhow to build large hydropower schemes, she will replicate them, both at home and internationally. A 2012 report from International Rivers identified Chinese involvement in 308 dam projects in 70 countries.

While Three Gorges is the world's biggest hydroelectric system, a larger one is proposed at the Inga Falls in the Democratic Republic of Congo. The Grand Inga Dam is planned to have a capacity of 39,000 megawatts, but was still at the early planning stage in 2014, when the World Bank rated Congo among the world's ten most difficult places to do business. The future of this project seems doubtful.

Enough electricity to run a street full of lamps can

be generated by the steps of a lot of people walking along a busy footpath or on a dance floor— pressure sensitive 'kinetic' tiles do this. These are in use in many places now, but will need to be a lot cheaper for general use. Power can come from use of the temperature differences between surface ocean water and the deeps, the movement of waves and tides, and volcanic activity.

The energy coming from the sun is unlimited in practical terms – its light and warmth nurtures all the planet's life forms, drives climate, and yet is available in such profusion that the sunlight falling on the earth for less than an hour could meet current world energy use for years. The trick is to find ways of harnessing it that are practical and economic. So far we haven't done so well with this. The 'old' hydrocarbons, coal, oil and natural gas, still provide 80 per cent of the world's energy. Crude oil accounts for 36 per cent, coal 23 per cent and natural gas 21 per cent. Most of the remainder comes from hydroelectric or nuclear powerhouses, and 'natural combustibles' like wood and animal dung.

The alternatives, things essentially driven by the sun like solar panels, wind, wave and tidal energy, make up barely three per cent. The use of these alternatives is growing quickly, but from a very low base. At the present rate of progress there is no way renewables could be relied on to replace hydrocarbon power fast enough to check climate change. Then there are the hundreds of millions of people who don't have any electricity. In 2015 the high-level Africa Progress panel came up with a ground-breaking idea to get power to them. The suggestion is that Africa bypass fossil fuel generation altogether, and that a $20 billion fund be set up to provide low-cost solar panels to the two thirds of Africans who don't have electricity.

Bob Geldof, who is one of the ten people on the panel, commented: 'Energy is the single most important key to eliminating poverty. Without power, you don't have the agriculture that's required. You don't have the economic development that's required. You don't have the education or the health that you need.' The panel noted that 600,000 child deaths a year were caused by household pollution from charcoal and wood fires.

A recital of megawatts of power produced by various means often doesn't mean much to most people, nor do I want to perpetuate the myth that one megawatt of generating capacity can provide power to 1000 modern houses in a developed nation. Even so, households are easier to envisage, so I will use an arbitrary but more realistic 500 houses a megawatt in what follows.

New Energy Algeria, a government-owned company, claims that country alone could produce four times present world energy use with solar thermal technology, in which large areas of mirrors capture and concentrate the sun's heat to boil water into steam. This drives turbines and generators, just as a coal-fired or nuclear power station does. Solar thermal power stations can equal or exceed the production of coal or nuclear plants, without any pollution, waste, or security problems. With the sophisticated heat storage systems now available, using molten salt in large vats, solar thermal can produce power continuously 24 hours a day. Spain's Gemasolar 'baseload' plant did this reliably over three years to 2014, supplying more than 25,000 households. It was the world's first major solar thermal generator to use modern heat storage technology.

There are quite a few others already up and running or in construction. The largest, the Ivanpah facility in the American Mojave Desert, can power 140,000 houses. A national programme launched in Algeria in 2011 provides for 40 per cent of domestic demand to be met by renewables by 2030 –that country has spent $315 million on an initial hybrid solar/natural gas plant — 'hybrid' means that solar energy will be used by day, with a natural gas supplement at night, when demand is low. Morocco is building solar thermal installations designed to serve 250,000 homes, providing 18 per cent of the nation's power, and at least 30 more are under construction or committed around the world. Saudi Arabia plans to spend over $100billion on the technology. The big Abengoa desalination plant in that country, designed to produce 60,000 cubic metres of water a day, will be solar powered.

There seems to be an immediate problem: how to get power generated in the desert areas of high sunlight to where it is needed – a transmission grid capable of carrying power over huge distances with minimal 'line loss.' If cheap solar power from the North African deserts is to reach Europe, it must travel under water, the Mediterranean. There is an established technology that meets both these requirements – high voltage direct current, rather than the alternating current in almost universal use now.

When Thomas Edison opened the world's first power station in Pearl Street, New York, in 1882 it generated direct current (DC), just what you get from a battery. However, at that time DC power could not be sent long distances over transmission lines at high voltages, then supplied to households at the lower voltages suited to their appliances. Rival tycoon George Westinghouse solved this problem with alternating current (AC), for which reducing transformers were available. After this came into use at the big Niagara Falls power station in 1895 it became the preferred technology worldwide.

However, there are problems. AC current tends to have substantial 'line loss' resulting from resistance in the cables – also AC cabling cannot run under water, as it would need to do to connect offshore wind farms. At very high voltages – as much as a million volts – direct current has significantly less line loss and can run under water. Line loss over 600 miles is less than three per cent, compared with ten per cent for AC lines. The longest link so far runs 800 miles from the Inga dams on the Congo River to the huge Shaba copper mines, which produce about 300,000 tons a year. There are comparable ones in China that carry hydroelectric power from Yunnan Province to the industrial heartland in the east.

China intends to triple its HVDC lines over the next few years.

A European model was to have been provided by a scheme called DESERTEC, using the perpetual sunlight and strong winds of the Middle East deserts to supply electric power and desalinated water to Europe. However the proposal was virtually dead in 2014 after it was abandoned by 16 of its 19 shareholders – chaos in the Middle East and its formidable cost of $480billion were said to be the major blockpoints, with most of the European economies in recession. A British offshore wind-farm company, Mainstream Renewable Power, is planning HVDC links between Germany, Scandinavia and Britain, and visualises an eventual 'super-grid' to include Europe and North Africa, at a cost of $275billion.

Wind-farms have more than quadrupled in capacity in less than a decade and in 2014 produced enough power for 160 million households in 83 countries. Almost a third were in China, the world's second largest user of wind power after the United States. Wind turbines are among the structural giants of the modern world, looming as much as 350 feet above the surrounding landscape, each capable of serving as many as 2000 households. Rotor diameters have almost doubled since the 1990s. The huge blade size of modern turbines allows them to work in lower wind speeds – blades as long as 300 feet are now in the development and testing stages, for use in turbines as high as a 40 storey building.

Wind power has reasonably been criticized as erratic – a turbine can only produce power while the wind is blowing. Plainly, if the alternate technologies are to produce a reliable supply, they must operate together – on a sunny day there may not be much wind, leaving solar thermal to carry the load, while on a breezy night wind farms will take over. This need for several technologies to operate together is a major argument for converting power grids to high voltage direct current.

A big research effort is being made for more efficient and larger capacity storage batteries, especially a type which charges the battery's electrolyte, or fluid. Theoretically, there is no limit to the size these 'flow' batteries can be built – those now existing are very large anyway – so they are being developed as a storage medium for alternate energy generators like wind farms.

The movements of the sea – tides and waves – and also the temperature difference between the deeps and the ocean surface are being researched and used to generate power. The world's biggest tidal power plant is in South Korea. After seven years under construction the Sihwa Lake station began powering more than 100,000 homes in 2011. It is rivalled only by a 50 year old tidal power plant at the Rance, in France, now the world's second largest. A Canadian power plant at Annapolis, in the Bay of Fundy, generates a much smaller 20 megawatts. Because the bay has some of the highest tides in the world, its potential energy output is very considerable – at least 5000 megawatts. In 2014 the

Nova Scotia government announced $4million in funding for further research into tidal power.

Britain is building six tidal lagoons at a cost of $50 billion which will power almost a million homes. In these seawalls ensure that water from the rising and falling tide must pass through the hundreds of generators set in the wall, providing power for 14 hours a day. Since these installations are designed to last for 120 years it is considered they will produce cheaper power than the proposed new British nuclear stations at Hinkley Point.

Geothermal energy, using hot springs or volcanic steam, is in use on a small scale in many countries, including the US, Japan, New Zealand and Iceland, to generate electricity and heat buildings. There are areas of hot subsurface rocks in many parts of the world, and several countries are researching ways to use this heat for power generation. Wave power can be harnessed, but most of the generators constructed so far have been small, expensive and subject to major damage in storm conditions. After a number of less than successful attempts, no commercial scale generator existed in 2015.

The energy problem is essentially quite simple and also quite difficult, requiring as it does a choice between three major sources — the hydrocarbons, nuclear and the renewables. Use of all three of these is possible — the degree of risk they involve should be the determinant of which and to what extent.

If we go on burning the hydrocarbons, coal, oil and gas, as our main energy source, we have been warned that extreme climate change with all its dangerous implications will become inevitable and ongoing for centuries.

Traditional fission nuclear power should not be a main option because of the risk of more nuclear 'incidents'. Even after 30 years Chernobyl still presents serious problems — it is costing billions of dollars just to control its radiation, much less get rid of the dangerous cause of this. The three melted-down reactors at Fukushima are so radioactive they can't be approached, and nobody knows how or when they can be rendered harmless. And in both places there remains an ever-present long-lasting risk of another out of control nuclear accident.

Because solar, wind and tidal lagoon generators don't present any risks, they pretty much select themselves as the preferred technologies — indeed the only safe ones available to us. Costs are falling anyway – – in many parts of the world the alternates are already providing cheaper power than that from traditional grids. We know how to make them, including how to deal with peak demand, so we should simply go ahead and do it.

However, even given a major increase in capacity, it would take many decades for wind and solar to provide as much energy as we use today. This is why energy is likely to be in short supply from about 2020 on. If we delay the transition until we are forced to it by intractable problems with the other technologies, the energy drought will become severe. If we want to avoid this, the shift needs to be massive, and soon. This means large segments of the economy mow making unessential and luxury products need to be diverted to the alternates as soon as possible.

This will be easier if we use less power, and do this painlessly. Thousands of 'passive' houses fitted with solar panels and water heaters now exist or are being built. Many of these produce more power than they consume, selling surplus solar production back into the grid. This must become the norm. The world is reducing its use of power, but not quickly enough. A study produced by the UN environmental Programme in 2015 showed that in 2014 the industrialized countries used almost 1 per cent less power than in the previous year, and slightly less than in 2007. It shouldn't be hard to do much better. Lighting, heating and cooling buildings uses almost a third of all energy, and the means to remedy this, like LED lighting, are available and pay for themselves very quickly. Transport of all kinds will need modifying to non-polluting forms. This is covered in chapter 20. Waste disposal methods that generate electricity should become universal.

8 The Nuclear Conundrum

The world's 440 nuclear power stations, which generate 16 per cent of its electricity, provide large amounts of clean power with minimal greenhouse emissions, but also present serious problems. While the damage from most disasters, like floods and droughts, can be fixed in a matter of years, the consequences of nuclear catastrophes persist indefinitely. When reactor fuel melts down, as happened with three power generators at Fukushima in 2011, there is no known way of dealing with them in the medium to long term. Those three damaged reactor cores are still emitting radiation so strong it would kill any human who approached them in a matter of hours, and they will go on doing so into the distant future. Even especially hardened robots sent in to observe them were disabled by this intense radiation.

The molten fuel can be controlled temporarily only by pouring large quantities of cooling water over it. This water, which becomes radioactive, cannot be stored indefinitely, and is spilling into the Pacific Ocean, where its polluting influence, at present slight to moderate, will increase with the years. There are also huge amounts of contaminated soil for which safe storage cannot be found.

Of especial concern is a major by-product of nuclear power, plutonium, which is a deadly poison and cause of cancer and remains dangerous for thousands of years. Everything possible should be done to keep it out of the air —a single small flake of it, lodged in someone's lung, can cause cancer.

Nuclear power is very expensive. Britain's proposed new reactors at Hinkley Point were estimated to cost $37 billion in 2015 for 3200 megawatts of power, and to meet this cost the British government has agreed to guarantee a price more than double current electricity market prices for these for 35 years, to be funded by levies on all consumers. However, that price estimate is speculative. Two plants being built in Finland and France are years overdue for completion, and will cost up to four times the original estimates.

While operating costs of a working reactor are low, there is another massive bill when they need to be decommissioned, and large amounts of radioactive waste disposed of. Britain has 20 aging reactors, which are due for decommissioning at a cost estimated at $400 million to $1 billion for each installation – in some cases more than they cost to build in the 1960s. The French Super Phenix fast breeder closed down in 1998 has already cost billions to dismantle, with continuing problems about how to dispose of 5500 tons of radioactive sodium, which would explode if exposed to water or air. Fifteen tons of plutonium are stored next to it. It would require a casing of 70,000 tons of concrete blocks to make the sodium safe. Meanwhile it cannot be allowed to solidify, and must be heated constantly to keep it liquid.

Promotion of 'fast' nuclear reactors as a future energy source seems less than responsible when so many have failed or have been closed as too dangerous.

The world does not have the technical infrastructure and expertise to build new nuclear power plants quickly — certainly not quickly enough to have any influence on the oncoming climate problems, because the specialized workforce required for manufacturing reactors has declined, along with the industrial base.

Although billions of dollars have been spent trying to find ways of storing nuclear waste, no satisfactory method has emerged, and most of this radioactive material is still in temporary storage. This is appallingly dangerous, as the continuing battle with the hazards of stored spent fuel at Fukushima shows – one of the largest and most dangerous accretions of nuclear material on the planet is held in damaged ponds above the wrecked reactors. During the many years it must take to deal with this, the ever-present danger of a violent nuclear incident within that fuel will continue.

A very dry part of Nevada in the United States, Yucca Mountain, had been proposed as a permanent nuclear waste repository, but in spite of exploratory work over 20 years costing $13.5 billion there were still problems that led President Obama to cancel the project in 2009. Residents of cities and towns through which connecting roads and railways passed objected to the facility strenuously, fearing a nuclear accident involving the waste while it was being transported. This transport problem is common to most nuclear waste disposal proposals. Other proposals, such as vitrifying the waste – enclosing it in glass cubes, then burying it — have proved too expensive. Shooting it off into space, an idea that has sometimes been canvassed, would be horribly dangerous if a rocket had to be aborted and the waste fell back to earth. Those proposing such things seem to have no idea of how much high level waste like spent fuel rods there is – more than 70,000 tons in the United States alone.

While on the whole nuclear power plants have had a good safety record, there have been exceptions that demonstrate what a huge risk malfunction can be if it does happen. Deaths from the nuclear accident at Chernobyl in 1986 are conservatively estimated at 4000, mostly from cancers. More than 300,000 people had to be evacuated from the contaminated area, including the entire population of the city of Pripyat, which is still deserted.

More than 20 years later, in 2010, forest fires burned into 2000 acres believed to be still contaminated, raising fears that radioactive particles, including plutonium, would be lifted into the air from the soil with smoke. There was another fire crisis in 2015. This is a continuing risk in the pine forest that surrounds Chernobyl, which cannot be properly maintained because of high levels of radiation within it.

The exploded Number Four Reactor is being covered with a huge metal canopy at a cost of more than $2 billion, to which 40 countries have contributed. It will have to be moved on rails almost half a mile to a position over the reactor, because of the intense radiation – in fact it is said to be the biggest moving land structure ever built. Its completion would allow work to start on dealing with the radioactive material still inside, possibly in 50 years' time, although at present no-one knows how to do it. It is salutary to note that radiation levels above 1000 roentgens an hour have been recorded on the site – 500 roentgens over five hours is a lethal dose, the average person's exposure in a normal environment is half a roentgen, at most.

The fuel for nuclear weapons comes from fission reactors that produce power. Only a small fraction of the uranium in a reactor's fuel rods is actually used to produce heat, and then power, while the rest is converted into a nasty assortment of other radioactive substances, mainly PU239, plutonium, which, as we have seen, is a principal ingredient of nuclear bombs. Nations who claim they want nuclear powerhouses only to make electricity also get the option of making a nuclear arsenal. They often seem only too willing to take this up, even though proliferation must greatly increase the risk of a 'limited' nuclear war.

High quality uranium is by no means a sustainable resource, and would be exhausted in a matter of decades if nuclear power were called on to be a total replacement for coal. Mining lower grade ores is a considerable source of greenhouse gas emissions. There is an additional cost in protecting nuclear power plants and waste repositories from attacks by terrorists who want nuclear materials, and from   aerial bombing in the event of war.

When all these things are taken into account the real cost of nuclear power becomes more expensive than that from wind energy. Because they can provide bomb fuel, existing nuclear powerhouses have been heavily subsidized by governments — this makes the apparent cost of power from them seem much cheaper than it actually is.

However, another way of generating nuclear power is being actively researched, especially by China and India. Thorium, an element that is around four times more common than uranium in the earth's crust, can be transmuted into a fissile isotope, uranium 233, which can then fuel a reactor. The fuel, which is conveyed into the reactor core as a molten fluoride salt, can reach temperatures as high as 700 degrees C – this heat is then used to generate electricity. There are some complications — the conversion of thorium into uranium also creates a very powerful but short-lived emitter of dangerous gamma radiation, thallium 208. This circumstance is said to make it virtually impossible to develop clandestine nuclear weapons with a thorium reactor, thus allaying the fears of proliferation that have been so closely associated with conventional reactors.

Thorium reactors are claimed to be safer than present reactors, and to produce less, and 'safer', waste, mostly caesium 137 and strontium 90, which are said to become virtually harmless in about 300 years. There is little doubt thorium reactors will come into use eventually. In 2011 the Chinese government allocated $350million to thorium development, planning to have a 2 megawatt prototype working by 2015. However, this has been extended to 2020, due to 'significant design challenges.' India has been working in this area for decades, plans to have a thorium reactor working by 2021, and hopes to supply a quarter of the country's electricity from thorium reactors eventually.

Then there is nuclear fusion. Basically, fission releases energy as atomic nuclei are split, while fusion operates much as the sun does, persuading hydrogen isotope ions – deuterium and tritium – to fuse, producing helium. This can only take place in plasma at very high temperatures, around 250 millionC. No physical object, not even the hardest of metals, could withstand such a temperature, so the process is confined in a *tokamak,* a complex system of electrical coils that creates a 'magnetic bottle.' Plasma is one of the four natural states of matter, the others being solid, liquid and gaseous – plasma in the sun and stars is the most common form of matter in the universe.

The world's most advanced fusion test reactor is now being built at Cadarache, in southern France. ITER, the International Thermonuclear Experimental Reactor, is the latest result of more than $60 billion so far spent on this technology, and is jointly financed by the United States, Europe, Russia, China, India and South Korea at an estimated cost of $12 billion in 2006. So far experiments into fusion have not reached 'break even point', at which more energy is produced than the process consumes, and the reaction has been sustained only momentarily. However, it is hoped that if fusion reactors eventually become feasible they will produce much more power than fission devices, with lower radiation risks and much less waste. Fuel for them is virtually inexhaustible and their sponsors regard them as environmentally benign.

Not everyone agrees it is a good idea. There has been a good deal of opposition from Greenpeace and other French environmental groups, who say ITER is costly and dangerous, and will never deliver any useful power. Enormous technical difficulties have yet to be overcome, with fears that intense neutron bombardment will damage the reactor structure enclosing the *tokamak*, that the confining magnets might be damaged, and even that the entire reactor may become radioactive. Unfortunately, by 2015 the project had run into that eternal problem of nuclear engineering, runaway cost blowouts, with construction costs alone estimated at $21 billion at least. ITER was initially expected to start running in 2016 – that has now been extended to 2025.

The next 30 years will decide whether these concerns can be resolved. However, whatever promise it might hold for the future, fusion power is unlikely to be developed soon enough to influence the oncoming problems of global warming.

A good deal of technical detail has been necessary

in these two chapters. I've tried to keep this to a minimum – although at the expense of more general statements than I would have liked. However, this matter of energy is a big deal for all of us, and the close association of today's energy sources with global warming makes it even more so. It is also an area where the efforts of individuals are of value. Everyone who puts photovoltaic panels or a solar hot water system on their roof, or agitates for a faster and more complete transition to clean energy, is working towards a necessary if at times painful transition. It is probable that the oncoming problems with climate and pollution will force the issue – this is already happening in the smog-plagued cities of China. Governments and people will have to cut down coal and oil use regardless of the economic consequences as this pollution becomes intolerable.

New technology, of course, can take several decades to get into place. Hence it seems likely the world will face energy shortages from about 2020 onwards, and that these will continue until new clean and sustainable sources are operating. Becoming self-reliant for power accordingly becomes an important element in the survival strategy of individuals. Solar panels, small wind generators and efficient battery storage, all of which are available, are obvious first suggestions, but there are others. People who are lucky enough to have a small stream running through their land can harness it – generating turbines that operate submerged are on offer at reasonable prices. Conservation, especially that provided by solar passive elements in properly designed houses, is a critical element – more on this in chapter 24.

9 Too Many People?

There has been a comforting idea around, sanctioned by some demographers, that world population will stop increasing at a manageable level late this century, and even decrease after that. Surely if this is so the problem of human numbers would largely solve itself – it won't get to being standing room only. Since it was taken for granted that population growth was no longer a worry, this issue has largely fallen off the human agenda. Now is the time to put it back on — the latest estimates from the United Nations not only make it plain that those earlier figures were wrong, but that we may even be dangerously close to the 'carrying capacity' of the planet – the most people who could be accommodated without serious and permanent damage to the global food supply and the environment.

That earlier assumption on population growth was based on birth rate estimates lower than what is now happening. Even if the two billion women now of child-bearing age have many fewer children than their mothers, global population will continue to rise for several decades. The crunch point is just how many they will have, and, hence, how much that total increase will be. Saying 'there was plenty of room for error in population projections', in 2012 the United Nations revised its estimate of 8.9 billion in 2050 up to 9.6 billion, with 10.9 billion by 2100, because birth rates in the industrialized nations and sub-Saharan Africa were higher than had been anticipated. In 2015 the world body raised the estimate for 2050 to 9.7 billion.

Late in 2014 an international research team led by Professor Adrian Raftery, of the University of Washington, said there was a 70 per cent chance world population would reach 11 billion by 2100. He commented: 'There is now a strong argument that population should return to the top of the international agenda.' Falls in fertility earlier predicted for Africa had not materialized, with the average woman living in the continent's largest country, Nigeria, bearing six children. If this continues Nigeria would grow from 200 million today to almost a billion at the end of the century, while Africa's population could rise from one billion in 2014 to 3.5 billion, possibly even 5 billion, by the end of the century. Professor Raftery remarked that the preference for large families was linked to a lack of female education, which limited women's life choices, and too low an availability of contraceptives. And, the most frightening prediction of all, these demographers say that if present fertility rates persist, there is no guarantee that world population will stabilize at 11 billion in 2100 – it is more likely to go on increasing. That is, of course, if we don't do anything effective about it.

The earth was sparsely populated until quite recently, its human numbers forced down by wars, disease at plague proportions and high rates of infant mortality. In 1350 a global population of 370 million, depleted by the European great famine and the Black Death, grew slowly to two billion nearly six centuries later, in 1926. From then on population growth has been much faster. There were three billion people 34 years later then four billion in not much over another decade, setting an alarming rate of increase that has persisted. It resulted in 7.33 billion people on the planet in 2015, and a much higher rate of urbanization. China is planning to merge 11 existing urban hubs in its southern industrial heartland in the Pearl River delta into the world's largest megapolis – 80 million people living in a region twice the size of Wales and with almost four times the Australian population. This project, which will include Hong Kong and Macau, is to cost $304 billion. However, the region is very prone to extreme weather events, and because so much of it is low-lying, likely to be affected by sea level rise.

Figures slip easily past the eye, but these projections represent something that has never been approached in history. The planet has grown no larger in the last century, but the presence of four times as many people has caused an immense shock to the environment, food-producing resources and water availability. In not much more than a decade, by 2025, two thirds of the world's population will not have enough water for their needs, according to the UN Environment Programme. Not long after that the world's fisheries will be depleted. Even now we each have half as much land to produce our food than was the case fifty years ago.

Just what is the 'carrying capacity' of the earth – the maximum population it could sustain indefinitely? Many scientists are working on this, and their findings, which vary considerably, certainly give cause for alarm. A United Nations survey in 2001 found that most of the estimates were between four and 16 billion, with a median figure of 10 billion, which is below current population estimates for 2100. Even with the minimal amount of nourishment, just enough to sustain life and health, and assuming most people became vegetarians, 10 billion seems as much as the planet could tolerate. There have been a number of estimates that if everyone ate as Americans, Australians and Europeans do, the earth could only support three billion. This is because of the large amount of food fed to livestock to produce meat – around 40 per cent of all grain production in the United States, for instance.

'Net primary production' is the total global amount of plant material deriving from solar energy converted by photosynthesis. The amount of this used by humans is our 'ecological footprint.' According to some ecologists we are in what is called 'overshoot' mode now, consuming resources faster than they can be restored – the Global Footprint Network estimated that in 2014 we 'exhausted earth's budget for that year in less than eight months.' Some demographers, like Robert Engelman of Worldwatch, believe we are already past carrying capacity. Failing water supplies and the effects of climate change can only make the situation worse.

According to a paper by biologist P.J. Bryant, of the University of California, prior to human impact the net primary productivity of the planet was about 150 billion tons of organic matter a year. However, 'by deforestation and other forms of destruction of vegetation, humans have destroyed about 12 per cent of this, and now use an additional 27 per cent'– for food, construction and other purposes. This means we have already taken over 40 per cent of the terrestrial food supply. Bryant deduces, on the bare statistics, that if world population reached 15 billion, it would take *all the plant growth in the world* simply to support the human presence. However he adds that when other factors are taken into account, such as waste and the need to preserve the eco-systems, 'the predicted carrying capacity of humans is much less than 15 billion; in fact, probably less than the current population.'

Bryant makes another bleak assessment: 'In many developing countries the population will probably stabilize not because of a decrease in the birth rate, but a return to higher deathrates, and this will reflect mainly an increase in the number of children dying from starvation-related causes.' (Peter J. Bryant. Hypertext book, Biodiversity and Conservation, 2005) As things are now, 15,000 children die from starvation-related causes every day. The professor of ecology and agriculture at Cornell University, David Pimentel, estimates the carrying capacity of the land in the United States at around 200 million – the population in 2015 was 320 million. In 2009 Professor John Beddington, the British government's chief scientific adviser, warned of a global 'perfect storm' of food shortages, falling energy reserves and population growth by 2030.

'Since the last ice age, nothing has been more

disruptive to the planet's ecosystems than agriculture…the global community now faces a crisis in land use and agriculture that could undermine the health, security and sustainability of our civilization.' These statements, by Jonathan Foley, director of the environment institute, University of Minnesota in an article in Yale University's Environment 360, are broadly typical of the views of many specialists in the field. The argument continues: 'We are putting tremendous pressure on the world's resources… If we want any hope of keeping up with these demands, we'll need to double, perhaps triple, the agricultural production of the planet in the next 30 to 40 years.' Producing beef and sheep meat uses so much water and protein and creates so much greenhouse gas its sustainability is doubtful, although some scientists say it might continue if people reduced what they ate to about half present levels. Raising chickens for meat has a greenhouse load around a quarter of that for beef.

India, with 18 per cent of the world's population, has little more than two per cent of the world's land area and less than two per cent of its forests. Around half of its people are living below the poverty line, many agricultural and industrial practices are not sustainable and there is relatively little scope for further expansion. India has had only limited success in controlling its population, and this has resulted in destructive changes in land use, increases in toxic chemicals to the environment and the depletion of natural resources.

Such comments could be applied to any of the

south Asian countries. The Himalayan state of Nepal is now the poorest country outside India, due almost entirely to population growth and the over-use of almost all of its cultivable land – a situation greatly aggravated by two disastrous earthquakes in 2015. Population doubling virtually every ten years produced 30 million Nepalese by 2014. More than 40 per cent are under 15 and the median age is 21.6 years. This situation is fairly typical of the world – large, fast increasing populations of young people in developing areas, compared with aging, often falling populations in the affluent countries.

Growth is slowing, but not fast enough. A peak world growth rate in 1963 of 2.2 per cent implied population doubling in 35 years; in 2013 it was down to 1.14 per cent, which nevertheless, if it continued, would double world population to a catastrophic 15 billion in 61 years. Birth control policies a number of world governments have instituted have slowed the rate of growth. The most celebrated of these is in China, where a 'one child' policy has been enforced by a number of draconian and unpopular measures, like enforced abortions and reduced living standards imposed on those who have unauthorised children. Fertility restraint by force is one obvious option to keep population growth down, but is it the best? Public opinion in China has been resentfully against it, and as a result the government has been forced to liberalize it by now allowing families to have two children.

Even in south Asia, where population increase is among the most problematic in the world, there are regions where growth has been controlled by modest prosperity and high literacy rates. Control measures, which include often unpopular vasectomies, have got the overall Indian fertility rate down to 2.6 children per woman, but in the crowded states to the south of Delhi – Bihar, Uttar Pradesh, Rajasthan, it is typically as high as four. More than half the women in this region are illiterate. They marry young, and live in villages where social status is linked to the number of children they have. In the southern Indian state of Kerala, 90 per cent of young people go to school — girls who are educated marry later, have a better understanding of contraceptive methods, and have fewer children. The fertility rate is actually below replacement, at around 1.7.

Rapid population growth was a problem in Japan, Singapore and Taiwan when they were less than prosperous and healthy societies, now they are these things their populations are stable or actually reducing. The transforming factor has been greater prosperity, a change from relative poverty to at least a modest and stable degree of wealth and security. This undoubted link between improved societies and the fertility rate was first put to me 40 years ago by a friend now long dead, Thomas Stapleton, who was Professor of Child Health at Sydney University, secretary of the World Paediatric Association and a recipient of China's highest award bestowed on foreigners, the Friendship Award. I can recall him discussing the issue with me with great vehemence, and there is no doubt subsequent events have proved him right.

Hence it seems evident there is a way to check world population growth. A major effort to eliminate poverty and disease through significantly increased foreign aid would not only be reasonable and compassionate, but would also result in large reductions in fertility –and might indeed be the best way to do this. Can we afford it? On the more moderate estimates the abortive Iraq and Afghanistan wars cost at least eighteen times what the world spends on foreign aid in a year. The wars cost $3.2 trillion (Brown University's Cost of War study), although other studies including such things as the cost of caring for returned veterans put it as high as $6 trillion. Global foreign aid amounts to about $170billion a year, and this includes money raised by non-government agencies as well as that donated by governments. Almost $1.8trillion a year is spent on weapons and other defence costs — this is around 11 times the aid budget. So the money is there all right, it's just a matter of whether we spend it usefully or waste it.

If we really want to hold down world population we need a major, much larger campaign to eliminate poverty, country by country, associated with programmes to promote birth control, better education of women, and a proper understanding by Moslem, Hindu and Catholic religious leaders of how dangerous uncontrolled population growth is. Conferences to achieve these things would need to be strictly practical – there have been innumerable talkfests in the past that have fallen short of the goal.

If solving one of the world's most fundamental

problems is judged more important than buying more weapons, why not provide this through a simple re-allocation from all defence budgets? These had inflated in 2015 to levels not seen since the Cold War. Since no nation would risk cutting its defence capacity unilaterally this would have to be negotiated worldwide, on the simple proposition that every nation transferred 10 per cent of their arms spending to a dedicated UN fund to eliminate poverty and improve health, water supply and other services in disadvantaged countries. If all contributed no nation would be at a strategic disadvantage – severe sanctions might be necessary to persuade the unwilling, such as military dictatorships, to co-operate. The powerful pressure groups that benefit from making weapons could transfer some of their capacity to infrastructure and commodities for the aid effort. Armed service personnel could be given the option of working in these programmes. This plan would double foreign aid immediately by making another $180billion a year available. Could this be negotiated, and how? -- the United Nations is there to do things like this.

The third necessary factor in this equation is educating girls. Educating girls? Surely they go to school like the boys? In places like Kerala, yes, but for a lot of girls—indeed on some estimates 66 million worldwide—no. Tens of millions more get a very inadequate education. Most simply help around the parental home, often caring for younger children, until they go to their husband's. Frequently this happens when they are barely pubescent. With no education, very little personal freedom and no knowledge of birth control methods they then have one child a year until they either die eventually in childbirth or from the stress of too many pregnancies. Instead of being in school like their Western counterparts, they often die young, and die miserably — according to the World Health Organization childbirth is the leading cause of death for girls aged 15 to 19 in developing countries. One in 160 15-year-old girls die in pregnancy and childbirth in those places, compared with one in 3700 in the developed world.

The effect these circumstances have on world population growth hardly needs spelling out. Since most poor people cannot afford school fees, books and uniforms, substantial outside assistance will be needed – a figure of $7.5 to 10 billion annually has been suggested. So all in all, it is fairly clear what needs doing, and that given money and dedicated expertize it can be done. Compared with the marvels our technology has been able to achieve, it should be no great deal.

10 Population and Poverty

Unchecked population growth and poverty feed on each other – this seemingly simple proposition is stated again because it is perhaps the most important single idea in this book. The demographic facts outlined in chapter 9 make it plain that crowding so many people on to the planet is not only uncomfortable but also dangerous – over-population is one of the larger hazards, and we don't have that much time to deal with it. This chapter looks closely at evidence that a massive global effort to reduce poverty may be the best way, perhaps the only way, to check runaway population growth.

To recapitulate, history has demonstrated again and again that as societies become more prosperous, their birth rate falls, and that when girls are educated, they have far fewer children than their unschooled sisters. If you have any interest in saving the planet, these may be the two most important facts to take on board.

Next, a tale – not of two cities but of a city-state and three nations. When I lived in Singapore 50 years ago, much of the city consisted of rows of overcrowded 'shop-houses' – shop below, living quarters above – threaded by a maze of narrow, noisome lanes, and closely settled *kampongs,* the Malay word for a village, which fringed Singapore in all directions. They were largely unplanned, just row on row of flimsy houses built of wood with thatched roofs they were little more than squalid slums –and they were an appalling fire risk. If one house took fire this would spread with frightening speed to its neighbour. I have seen the ferocity of one of these *kampong* fires, and it is something I will never forget. The worst of them came on a windy day in 1961 when more than 100 acres of Kampong Bukit Ho Swee were destroyed, making 16,000 people homeless.

It has been observed that modern Singapore was born out of fire, and this is probably true. From 1970 on a major redevelopment plan has transformed it into localities of modern high-rise housing within one of the world's most prosperous, innovative and healthy cities. In the days of the *kampongs* and crowded shop-houses a major problem was an alarming rate of population growth – 4.4 per cent annually, which, if continued, would have doubled the population in 16 years. A 'stop at two' campaign was firmly established, and was so successful that by 1975 the fertility rate was below replacement. It still is. In 2014 it was one of the lowest in the world –1.2 children born per woman, in spite of inducements like cash baby bonuses, with the population actually reducing.

Unchecked population growth was one of the major causes of one of the world's great tragedies, the involvement of Japan in World War Two. This involved the destruction of two cities, Hiroshima and Nagasaki, by atomic bombing, huge damage to 66 others by fire bombing, the death of almost half a million civilians and one and a half million combatants in Japan alone, plus major damage and loss of life in the fought-over areas of Asia and the Pacific. Japan's violent and cruel attempt at empire was driven by a need for food for her rapidly-growing population, which had increased from 55 million in 1920 to 73 million 20 years later. Korea and Manchuria, taken over before the war, were exploited as food sources and quite soon after war began Japan had conquered the prolific rice-growing regions of Indo-China and the oilfields of the Dutch East Indies.

Japan emerged from defeat greatly changed. Twenty years after the war she was a modern industrialized society, in which nearly 90 per cent consider themselves part of the middle class. The result? Japan's population in 2015,127 million, had dropped by almost a million in 5 years, and is continuing to fall, with 100 million projected for 2060. Well-educated and prosperous couples are increasingly disinclined to have children. A clumsy description of Japanese women by a health minister as 'birth-giving machines' aside, the government is only too aware of this problem, providing generous child allowances and subsidies to child care centres as incentives for parentage. Photographic equipment maker Canon closes its offices early – 5.30 pm – two days a week, so workers can go home and 'make babies.'

Taiwan has had a very similar experience for much the same reasons. The birth rate is well below replacement now, but in 1964 a fertility rate of 3.5 per cent threatened population doubling in 24 years.

Population trends in Bangladesh, a large, poor South Asian country, support the proposition that numbers can be controlled if enough considered effort is applied. Bangladesh, one of the most densely populated places on earth, has 153 million people – the population almost doubling over the last 40 years. Even though vigorous birth control measures are in place, it is predicted the population will rise to 235 million by 2050, then reach 320 million by the end of the century. There are complex reasons for this continued high growth in what is called 'the late transition stage', among them a falling death rate and higher intrinsic fertility because of better overall health.

That figure of 320 million is bad enough, but if it were to grow even more the results could be catastrophic. As it is the average Bangladeshi woman now has 2.15 children, only fractionally over replacement, compared with 5.1 in 1981. Bangladesh has an independent Directorate-general of Family Planning, and there has been a continuing programme of foreign aid costing $13 billion, much of it devoted to health and fertility issues.

Nearly 24,000 field workers, mostly women, have persuaded 60 per cent of the country's women to adopt contraceptive methods, in the words of one of them, Anwara, 'by winning women's trust.' This has happened in an Islamic country, where there has been considerable religious opposition to family planning. Given help with the means, nations with a population problem can put together effective family planning mechanisms.

The world is rich enough now to control and even eliminate poverty in a matter of decades, and to see that girls everywhere are properly educated. Trillions of dollars have been wasted on pointless and destructive wars while the fundamental needs of humanity are still largely neglected. Polluting, expensive and unnecessary planned obsolescence and excessive packaging in the West are almost as wasteful. Many people may have seen the videos circulating on the Internet of the millions of new cars in storage that can't be sold. Plainly, the world has far too many cars, and the wrong sorts of cars, but their makers go on churning them out to keep their production lines moving, although this involves a massive energy and financial cost. And it's not only cars — there's an excess of many other consumer items.

But throwing money at the problem is, of course, just a necessary beginning. Advice and assistance by trained people on the ground would also be needed to see that money was spent effectively. There are millions of unemployed young people in western countries, many of whom could be willing and able to undertake the training that would fit them for these objectives. Volunteer corps like this have worked effectively before, but never in enough numbers to do the job.

If we fail in this, what might happen? Among other things, if current world population trends continue, the peoples of the West will become an inconsiderable minority during this century. Between 1950 and 2000 the Caucasian population grew by two-thirds to 1.25 billion. In Asia, Africa and South America in that half century it grew more than two and a half times to 4.75 billion – effectively, Westerners dwindled from a little under half to less than a quarter of the world's total population. The 'more developed' nations are set to represent only 15 per cent by 2030 – the best estimates are for rather more than 7 billion in the developing world, fewer than 1.5 billion in the developed world.

The prediction of 18th century philosopher Thomas Malthus that humanity would over-populate the planet, causing a mass famine, has been discredited in the past, but now it is back with us, a very evident threat. In 1998 UN demographers set out three scenarios for future population growth. In a scary worst scenario, in which poverty in the developing countries continues to increase, 15 billion people before the end of the century was predicted, in spite of an appalling rate of infant mortality. In 2013 the UN's World Population Prospects raised its estimate for 2050 to 9.6 billion, 300 million more that it had previously forecast, because fertility levels in some places, such as Africa, had not declined as much as had previously been expected. In 1950 Africans made up ten per cent of the human race. On present rates of increase that will grow to a quarter of humanity by 2050 – 2.7 billion people. A total world population of 11 billion seems likely for 2100 — the only region where numbers will fall is Europe.

The need to improve the lot of the poor on compassionate grounds is obvious, but the catastrophic demographic, economic, environmental and social consequences if the remedies are delayed are less so. This large problem breaks down into millions, of parts, the lives of individual people whose needs must be met at basic practical levels.

Close on half of all humans still live in villages – there are more than two million of these in India and China alone. In theory these villages are self-sufficient, but in reality they are becoming increasingly over-crowded and impoverished. One of my most uneasy memories is of a hamlet in North-east Thailand, housing about 150 people, an offshoot from a larger village. I shall call it Mae San. Young people were forced to move to this barren hillside because there was no longer room for them in their home village. It quickly became a place of the damned – that is a considered description. The soil was too poor to provide enough food — the young couples didn't know how to get fertilizer and couldn't afford it anyway. These people were slowly starving to death. Every child – and there were plenty of them – had the bulging 'pregnant' bellies that come from acute malnutrition. These are the people who have to kill their newborns because they know they can never feed them, or sell their daughters into prostitution while they are still children. Large numbers, and they count themselves the lucky ones, get to move into the city slums, where some poorly paid work might be available.

The needs of these people, for many decades at least, will be quite different from those of people in the Western world, and may seem simple to the point of being primitive. Nevertheless, those needs are real and urgent, and are frequently and disastrously misunderstood, even ignored, by decision-makers administering foreign aid. They prefer to throw billions at large capital works, rather than be bothered with the complicated problems of poor families.

A conventional view of history sees the end of the colonial era about halfway through the 20th century, when most of the former Asian, African and American colonies became independent states. However, this is only partly true. During the next half-century economic dominance of many of these new nations by the affluent nations has continued, partly due to deliberate exploitation, partly because many new nations seem incapable of to running their own affairs. 'Government' by military juntas and other autocracies has become common, and in too many cases these oppressors have been supported by Western powers for muddled political and economic reasons. This continues in spite of the regular failure of even well-intentioned aid programmes – in the crudest cases simply because aid funds have been diverted into the pockets of the 'leaders.' Other aid has been dissipated on inappropriate infrastructure like golf courses and city office buildings, and on weapons systems used by the juntas to oppress their own people.

Closely associated with all this is a financial racket that is killing thousands of children every day. These deaths are just one cost of the crippling burden of Third World debt, variously estimated at between $1.5 to 2 trillion. Paying interest on this, and trying to repay at least some of the capital, is frequently at the expense of proper health care, education and poverty alleviation in developing nations. Western countries and financial institutions have provided loans than often turn out to be 'odious' debt – money paid over to support tyrannies or, in some cases, debts inherited from colonial regimes.

The global financial crisis and the relentless progress of compound interest have worsened this growing problem of debt, leaving many poorer countries obliged to meet interest payments greater than their spending on education or health. For instance in Ecuador, whose generous natural resources ought to guarantee prosperity, two-thirds of the people live in poverty, largely due to interest payments that have amounted to as much as 40 per cent of the national budget. Paying interest rates as high as 21 per cent Ecuador has made debt repayments that exceed the principal borrowed. Of all loans made between 1989 and 2006 only 14 per cent was usefully employed. The remaining 86 per cent went to paying for previously accumulated debt. Millions of people don't like Third World debt – 17 million signed a petition organised by a non-government charity, Jubilee 2000, calling for it to be written off. Some token gestures followed, but the major problem remains. Jubilee 2000 clones in many parts of the world are still working on it.

A dispassionate look at this complex mess of poverty, population and debt makes it apparent that the world's governments and big banks are either incapable of doing much about it, or don't really want to – Australia, for instance made savage reductions to foreign aid in 2014, and again in 2015. Nor are they likely to perform any better in the future. This means that if anything is to be done, you and I will have to do it. Of course we already are. Dozens of NGOs – non-government organizations like OXFAM – around the world are doing practical work on the ground to help people in thousands of villages install reliable clean water supplies, improve livestock and food crops and develop new cottage industries. Almost without exception the NGOs are the most effective at getting maximum results per dollar spent and directing aid to the people who really need it. But even though they do their best, the problems are too vast for them to cope with.

Let us return to Mae San, the village of the starving described earlier in this chapter, and consider, in some detail, how it might be helped. There are already lots of small groups in Western countries that have 'adopted' poor communities, who raise money to help them and organize practical new infrastructure – perhaps a well for a clean water supply, books or stationery for a school, bed nets to keep out malaria-bearing mosquitos at night. Such groups, organised on a massive scale within 'parent organizations', experienced international charities, may well prove to be a 'breakthrough' mechanism. Those parent organizations could get together expert volunteer groups – perhaps made up of retired trades people — who would visit places like Mae San and report back on what needs doing, and in what priority. In some countries retired tradesmen are already doing this in a small way, but the need is for many more grouped into properly organized and adequately funded teams. Their reports would be made available to community groups in Western countries who are prepared to raise or donate money to help impoverished villages. There is evidence that people are more willing to do this if they can identify with a particular place or group of people. Let us say a group somewhere in north London decides to help Mae San. First up, they would get a disc with a little documentary about Mae San and its problems, so they get to know the place and can develop personal feelings about it. This tangible connection between one group and another greatly different one adds a human dimension generally lacking in officially organized aid programmes.

Because it is typical of thousands of others like it, Mae San needs pretty much everything, but some things more urgently than others. They are broadly these:

*The means to grow more and better food.

*The knowledge and means to limit the number of children.

*An adequate and safe water supply.

* Improved cooking stoves with chimneys so cancer causing smoke can be vented outside houses.

*Basic medical care and essential supplements like vitamin A tablets to prevent child blindness.

Given these things Mae San could be transformed from a death trap into a community with some dignity and some expectations. On this base a second level of development could be built – a school to be attended by girls as well as boys, solar-powered pumps to distribute water rather than back-breaking carting of water in buckets, roads and bridges, some cottage industries… this progressive list could go on until at last Internet access and simple computers were made available to the schools and to farmers. The need for fast, accurate information will become all the more necessary as these communities progress.

But for now it is worth homing in on our top priorities, why they are so important and what might be done about them.

Food: Typically, the available soil is worn-out, with very low fertility. Even after much hard work it grows only meagre crops with a low nutrient value. Because early results will save lives, in the short term this can be best be remedied by the application of fast-acting fertilizer and in the case of sour soils, lime or dolomite. Yes, compost, of course, as a longer-term solution, but let's keep the kids alive while it's making... Also right up front is a need for better agricultural hand tools, spades, rakes, hoes, preferably made of long-lasting stainless steel, and high-producing, tough food crops, including legumes, because these take nitrogen from the air and get it into the soil. Those with a high protein content should be preferred, because protein deficiency is arguably the nastiest form of starvation.

There is a wonder crop that not only fills all these requirements, but will also grow in hot dry climates and a wide variety of soils. Granted a reasonably high proportion of phosphates in your general fertilizer, the cow-pea could be producing food with a protein content up to 25 per cent in Mae San in quick time. Cow-peas are already cultivated in much of Asia and Africa. Drought-tolerant, they grow well on poor, sandy soils.

Even the dried stalks can be used as animal feed. High quality fruit trees and root crops also need to be provided and planted. Fast-growing 'living walls', hedges grown from cuttings, would protect these gardens from predators.

In Mae San trees tend to be cut down for firewood and building materials, and without intervention this is likely to continue. Maybe an answer would be to provide trees that produce food – fruit orchards – which the local people would see as valuable enough to conserve. The African baobab, with its massive trunk and feathery top, is one of the world's most extraordinary trees, and its fruit, not well known in much of the world, among the most nourishing. This 'super-fruit' contains three times as much vitamin C as oranges — twice as much calcium as milk, more iron than red meat and a large range of other minerals and anti-oxidants. Its juice is sweet but tangy, resembling pear juice. Baobabs can store up to 100,000 litres of water inside their trunks, making them suitable for harsh drought conditions. They are believed to live for up to 3000 years. Because baobab fruit is now in demand in the Western world, they could provide a valuable cash crop for thousands of villages.

Contraception. Those who for religious reasons oppose contraception in places like Mae San should reconsider their motives urgently. What sort of god would want millions of children to die horrible deaths in infancy – many by outright murder — if this could be avoided? Beyond this, it becomes very difficult to help these communities if they are swamped with hordes of hungry children. Somehow human fertility must be controlled. Teams of well-informed people should go to Mae San as soon as possible to offer the people there the facts, a large supply of condoms, and any other acceptable means to control fertility. If results elsewhere in the under developed world are a guide, the great majority of people will be only too happy to limit their families.

Water: Unhappily, this necessary substance is also one of the world's great killers. According to the World Health Organisation, four per cent of all disease globally is water-borne. Diarrheal infections alone kill 1.8 *million* people every year, mostly children – 5000 of them a day. This is because a billion people do not have access to clean water. Of course it isn't the water that kills, it is the myriad infections like cholera and typhoid carried in the only water available to those people. This is the case with Mae San – water has to be carried nearly a mile from the muddy infected river downstream from a dozen other villages. There is no other reliable water source nearby – if there were one, water could be piped from it, purified, and used for all drinking and cooking.

Mae San does not have this option. It does, however, have very heavy monsoon rain for some of the year, very little for the rest. There is no nearby valley suitable for damming. The only alternative then, seems to be collecting rain from roofs, but at present this cannot be done because almost always they are thatched. Thatched roofs don't last long, so replacing them is labour consuming. The answer could be to provide roofs that do last longer and which can deliver water efficiently into storage. They must also be cheap, light enough to be carried into villages served only by footpaths, and able to be installed by unskilled labour without specialist tools. Aluminium suggests itself since it is light and won't rust. Corrugated aluminium roofing could be carried in rolls. A simple metal hook that clipped over bamboo roof purlins, secured by a plastic washer and a wing-nut, would serve as a fastening. Preformed guttering could be attached in much the same way, and a pipe would allow gravity feed to a tank inside the house, large enough to provide drinking water for an extended time. All this could be done without special tools – a punch and a hammer could make the necessary holes in the sheeting. If systems like this were in place in all the villages with a water problem, and simple hygiene explained to the people living in them, it should be possible to break the deadly cycle of water-borne infection within a single generation.

Medical care: Western people take the availability of this for granted, but it is not readily available to most people in the world, and certainly not in Mae San. What is needed first is information. I found when working in North-east Thai villages that there was no general understanding of basic health knowledge – how mosquitoes spread disease, how dangerous contaminated water can be. This was at a time when hundreds of children were dying from haemorrhagic fever, the deadliest form of dengue, which is mosquito-borne. We corrected this and child deaths were greatly reduced with a dramatized radio serial using local dialects to explain basic health issues – in most villages perhaps three or four people had cheap transistor radios. Doctors speaking Bangkok Thai had previously lectured people, without success. However, the villagers were fascinated by the dramatic form and, as with people everywhere, once they knew the dangers, they acted to avoid them.

Trained 'barefoot' doctors, carrying the necessary drugs and instruments in a backpack to treat common ailments, could deal with many of the villagers' health problems. Each one could have a 'circuit' of perhaps five villages, visiting each one once a week. They would be able to refer cases beyond them to a district hospital. Of the world's almost 50 million blind people, 90 per cent live in poverty-stricken areas of Asia, Africa and Latin America. Eighty per cent of these could have kept their sight using established procedures. Almost all of the 20 million with cataracts could be cured for around $20 each – having had that miracle operation myself I am very happy to contribute to that cause. Our 'barefoot' doctors would certainly carry in with them cheap vitamin A tablets, which would prevent millions of children going blind at a cost of about 50 cents a child. Children blinded by vitamin A deficiency usually die, mostly from starvation, within two years.

In 2014 the World Health Organization published a shocking statistic, revealing that air pollution killed eight million people in 2012 –more than AIDS, road deaths and diabetes put together. More than 70 per cent of these deaths were in low and middle-income nations in Asia. While the dangers of severe air pollution in big cities is now well publicized, less has been said about the fact that slightly more than half these deaths were caused by indoor smoke from wood and coal fires. A mass-produced metal front, top and chimney designed to fit into a stove made from local brick or clay could solve that problem.

On present indications a massive reduction in human numbers from starvation and disease seems inevitable — the beginnings of this can be clearly perceived now. I have heard people say this will just happen, and what can we do about it anyway? — an excuse to do nothing. However, every human life is important, so we should be actively planning to keep that death toll as small as possible. There is nothing specially difficult or arcane about this. As these chapters have indicated, the means for social improvement are well known — and the world has quite enough money and unemployed young people who could be trained to do the job. The more villages, the more families, that can be lifted from poverty, the greater their chance of surviving whatever catastrophe we encounter. Accordingly, our major shift in aid policy should not be in five or ten years, but now.

Because the degree of need is so great, some thought needs to be given to guidelines for what must be a massive effort. Priority should be given to the places of most need, regardless of ethnic, religious or political considerations. This suggests independent instrumentalities should direct the programmes, rather than diplomatic bureaucracies, who have done so badly in this area in the past. The initial accent should be on enough food to avert starvation, provision of clean water and population control, and the direction should be towards individual villages, even individual families within them.

World population after the disaster must be controlled and its number regarded as the norm. The impact will be heaviest in the poorer countries, where the survivors will need a great deal of help. If they are left to struggle in poverty and ignorance they will have many children, and before long the planet will be over-populated again. This must not happen.

The huge aid effort required will need very large numbers of trained people. There is now a better understanding of the appeal daesh and similar organizations have for young people, and the extent to which the terrorists exploit their idealism. Youth has always been attracted to causes, and if it is denied these by its own society through failure to provide even the most menial employment, it will look elsewhere for them. Many of the millions of unemployed young people around the world would probably be willing to be recruited into youth corps to work with the poor and dispossessed, and should be organized to do so.

11 Food

At some time in your future you will probably consume food devised in the laboratory, even eat insects. You may not know you are doing this, but there is good sense to it. According to their proponents biscuits made from ground up European crickets are more nutritious than those using wheat flour, with a 70 per cent protein content, and are said to taste just like any other biscuits. They may even by flavoured, say with apples or honey, by feeding the insects these foods not long before they are 'processed.' According to researchers small ants have a pleasant lemony tang, giant water bugs taste like salty apples, termites are minty, dried grasshoppers contain 60 per cent protein, locusts are full of iron. These are some of the insects already used for food. There are even salt and vinegar-flavoured crickets available – although generally people baulk at cockroaches. Insects are cheap to rear and easy on the environment — according to a UN report it takes 10 pounds of feed to produce a pound of beef, while a pound of cricket requires less than two pounds. Farming insects produces barely one per cent of the greenhouse gases and uses much less water than conventional livestock.

So if we are going to feed our future billions, all our attitudes and prejudices about food will need to change – most of all, the ways we grow it and their relative efficiency. Rereading Jonathan Swift's *Gulliver's Travels* — not really a children's book, by the way – I came across these words: 'Whoever could make two ears of corn, or two blades of grass to grow upon a spot of ground where only one grew before; would deserve better of mankind, and do more essential service to his country, than the whole race of politicians put together.' This is probably truer now than it ever was.

Most food comes from dirt one way or another, and good quality dirt is something the world is running short of. The UN's Food and Agriculture Organization estimated in 2013 that   the degradation of productive land was costing $40billion every year – and this was a crude cost, not taking into account hidden extras like increased fertilizer use and loss of biodiversity. Over-use of the land for grazing, ploughing and deforestation is destructive, leaving the soil highly vulnerable to wind and water erosion –

– yes, of course, all a great shame but it's a less than perfect world, isn't it? … Indeed it is, but more importantly, have you ever been really hungry? Maybe at some time things caused you to miss meals for, say, twenty-four hours — try to remember how you felt, weak, cross, cranky, above all, empty? This chapter about the world's oncoming problems with food brings in this personal note because too often statistics – lines of figures – are run in front of us, but we don't really take in what they mean. Well then, try to visualize how you'd feel if you couldn't get food for, say, a week. You could do this and still stay alive, but after that week you'd find it difficult to do most of the things you normally do. You would start to get acute stomach pains and feel so

threatened you'd be afraid you were going to die. Your body and mind would combine in screaming out for food, any kind of food, anything you could get down into your stomach. Having visualized this you can begin to understand why starving children will eat grass, bark, even mud. This experience is not uncommon, many thousands will go through it, and of these the majority will eventually die from starvation. Try to forget about the politically correct euphemisms – under-nourishment, malnutrition.

No, starvation is the word — and it is dangerous, not only because of its lethal threat to individuals, but because of the way it can destroy established social orders. Famine is grist to the Steamroller, because morality, respect for law, for property and even for other individuals all yield to those overwhelming pangs of hunger. The starving will do almost anything to get food of any kind, cheat, steal, kill – there are even well documented cases where people from civilized societies have resorted to cannibalism.

Numbers? Around 800 million people don't get enough food — millions of the starving are already fighting to get it, or the land on which they might grow it. Forty-two per cent of Indian children are stunted because they don't get enough to eat – that amounts to at least a hundred million children in just one country. Although this is not the only cause, fighting over available land and water is a major element in the 12 year conflict in the Darfur region of the North African nation of Sudan. In 2015 Sudanese government sponsored militias were still murdering, raping and pillaging there, destroying 115 villages. In the village of Tamarawa nine women were cut down by machine gun fire when they tried to prevent militiamen stealing cows belonging to the villagers. According to Human Rights Watch almost half a million people have been displaced there since the beginning of 2014 – a USAID situation report puts those affected one way or another at almost five million people. Conflict, famine and disease have killed more than two million since hostilities began. While the World Food Programme (WFP) in 2009 provided emergency food for more than a million people, even this aid has been disrupted by regular attacks on and looting from UN food convoys.

That effort, costing more than a billion dollars, was just for Sudan. The WFP provides food assistance to more than 80 million people worldwide. That single figure is surely stark enough to show how huge the problem of hunger is getting to be. The cost of all this food aid, more than two and a half billion dollars a year, is met by donations from almost every nation. The lion's share, almost a billion dollars, comes from the United States, with Canada, Europe and the United Kingdom each paying around a quarter of a billion. The WFP, the largest human agency fighting hunger, was initially established in 1963 on a three-year experimental basis, but has had to be continued because millions would starve to death without it.

But to return to Sudan: this place of general

lawlessness and social breakdown is afflicted by an apparently never-ending conflict between two major groups competing for land and water. The villages of sedentary subsistence farmers, which tend to be grouped around sources of water, are being raided and their land usurped by nomadic tribesmen, mostly Arabs, who are competing with black African farmers for the same scarce water. Hungry groups raid cattle, which they sell to get wheat. Imposed on this pattern of social chaos is another group, armed bands of young Arabs known as the *janjaweed* – now rechristened the Rapid Support Force – who attack and burn villages, kill and rape women and children. They are supported and encouraged by the national government in the Sudanese capital, Khartoum.

Behind all this is the added influence of climate change. According to the World Environment Programme, there has been a 30 per cent drop in rainfall in the region over the last 40 years. Much of Darfur is arid anyway, but increasingly long spells of drought are turning more of the land into desert. Population pressures, over-cultivation and over-grazing are degrading it, so it provides increasingly smaller yields of food.

The Darfur situation is appalling in itself, but it is also a significant indicator of how bad things can get if similar competition for food and water develops in other parts of the world. Unless we want increasing chaos and millions more hungry refugees we have to stop this happening. Something new and different is needed. In the end this boils down to how we choose to spend money. Should we pour millions into conventional agricultural research in the hope of another 'green revolution,' or are there other ways?

It has been a popular perception that the world has plenty of food, but millions of people don't have the money to buy it. While the second half of this proposition is still true—the price of staples like meat, dairy and grains has at least doubled since 2000 — the first is becoming increasingly less so. It is too soon to say definitely that the massive droughts in China, Russia, Australia, Morocco and some other grain-growing areas have been due to climate change, although there is good evidence they were. World grain production per capita has dropped since a peak in 1985, in spite of record harvests in some years. The droughts reduced reserve stocks in 2013 to 423 million tons, enough to meet world demand for 68 days. This is coming uncomfortably close to the 64 day supply that fuelled a major spike in food prices and massive rioting in several countries in 2008.

Wheat prices spiked in 2010 as 160 year record heat waves, drought and wildfires reduced production so much in Eastern Europe that Russia suspended exports. This region normally supplies a quarter of the world's wheat. A bumper grain harvest in 2011 – almost 2.3 billion tons – did little to increase reserves because of increasing demand, while more good harvests in 2013 and 2014 brought a 'modest' increase in per capita food availability – 0.3 per cent.

According to an Oxfam report in 2013 'a hot world is a hungry world.' This report predicted that the cost and availability of food would be affected by increasingly severe weather. The agency estimated the average price of food, including staples, will double again in 20 years, in part due to erratic weather patterns and an acute drop in rainfall caused by shifting climates. Crop quality can also be expected to fall.

While world population is increasing, productivity is limited because there is little new cropland available. According to Cornell University academic David Pimentel per capita cropland is declining, and now stands at about half that needed for a typical 'rich world' diet to be provided to everyone. His estimate of an optimal human population is under two billion. Production in the main growing areas – the United States, Canada, Europe, Australia and Argentina, depends so heavily on declining oil stocks that enough production increases to match population growth seem unlikely. High productivity in these countries relies on agricultural machines, insecticides and fertilizer, all dependent on cheap oil.

This should give some insight into the nature of the trap waiting for us. How catastrophic could a resulting famine be? Some experts in this area spell out a chilling scenario of only enough food available to feed three billion people – this in a world in which by 2030 population is estimated to grow to more than eight billion. What will happen to the other five? The Pimentel view is 'the world must develop a plan to reduce the global population to about two billion. If humans do not control their numbers, nature will.'

Researchers at the Lawrence Livermore and Carnegie institutions estimated that global warming cost the world about $5 billion in reduced food output as early as the 20 years to 2002, with harvests of corn, wheat and barley 40 million tons a year below what they might otherwise have been. 'Most people tend to think of climate change as something that will impact on the future,' one of the authors of the study, Christopher Field, remarked, 'but this study shows that warming over the past two decades has already had real effects on global food supply.' In 2015 Dr Field said 'We're seeing slower than expected yield increases at a global scale... The impacts of changes that have already occurred are widespread and consequential.'

The IPCC's 2007 report said increasing temperatures and extreme weather patterns were reducing crop yields in heavily populated parts of Asia. Its predictions for the future were grim - 50 million more people at risk of hunger by 2020, 132 million by 2050, 266 million by 2080. In his discussion paper *The Coming Famine* Julian Cribb, of the Sydney University of Technology, says that unless we devise new ways of producing food, by 2050 three-quarters of the human population will be living in places where they are totally without the means or the knowledge to feed themselves. Cribb predicts 'our giant cities will be gigantic death traps, at the mercy of even minor glitches in regional or global food supplies.'

Several oil-rich or otherwise wealthy nations are already buying or leasing fertile farmland, mostly in developing countries, in a move to forestall future food shortages in their own countries. Half of the Democratic Republic of Congo's agricultural land has been leased out – one Brazilian and Japanese project growing soybeans and maize for export occupies land there bigger than Austria. Qatar has leased almost 100,000 acres in Kenya in a $2.3 billion deal. The land, which is in the fertile Tana River delta, will export fruit and nuts to Qatar. Kenya is already experiencing severe food shortages. Just a few rich families own vast amounts of farmland, while millions live in densely-packed urban slums. Saudi Arabia and the United Arab Emirates are negotiating leases for large areas of farmland in Senegal and Sudan. China's Fuhua Group plans to invest $4 billion to grow crops in the Philippines. A move by South Korea to lease almost half of Madagascar's arable land was forestalled by massive public protests. These moves to take over productive land must inevitably prompt uneasiness, with science fiction-like visions of thousands eventually starving outside the wire in their own countries.

Predictions for China, the world's most populous nation with 1.4 billion people, give a clear warning. China has 20 per cent of the world's people, but at best ten per cent of its arable land. Twenty-two of its 31 provinces were affected by drought in 2015, with huge areas of fertile farmland going out of production as they were overwhelmed with sand blowing in from the Gobi.

As early as 2007 the head of the State Meteorological Organisation, Zheng Guogang, warned of 'a severe impending crisis in China's food situation because of global warming.' He forecast that climate change would cause more severe droughts in already dry areas, with rainfall reduced ten to 30 per cent by 2030. This could cut yields of grain by as much as 50 million tons. China will have 200 million more people by then, requiring an additional 100 million tons of food. These ominous figures plainly do not add up. China, which had declared self-sufficiency in the 1990s, has had to import grain from 2010, five million tons of in 2011, while capricious weather struck again in 2013, ruining much of the grain harvest, and prompting opinion that China might soon become the world's largest grain importer.

Since there will be billions more people in the future world who have to be fed it is urgently necessary to consider ways this might be done. Here are some very broad categories:

*Reducing waste, by conserving the huge quantities of food either eaten by pests, overtaken by rotting in developing countries or thrown away in affluent communities.

*Developing improved land intensive farming methods resistant to climate change in the main food producing countries, and the promotion of new crops.

*Improved soil quality, the provision of better seeds and other growing stock, and more efficient farming methods in village societies in the developing world.

*Reduced use of grain as a livestock food to produce meat, and to make biofuels like ethanol.

*Manufacture of artificial substitutes for naturally grown food.

A British Institute of Health report claims that at least 30 per cent of all food bought in shops in the United Kingdom is thrown away uneaten. Enquiries in other parts of the rich world indicate a similar, if not worse, situation. This is a difficult issue. People feel that if they have the money to buy food they also have the discretion to either use it or throw it away. But do they, when so many people are going hungry? Food was rationed in wartime because there seemed a clear necessity to do this. It could be argued that clear necessity may well emerge again soon.

However, the truly massive waste is in developing countries like India. John Floros, Dean of Agriculture at Kansas State University estimates as much as half of all food harvested is lost before it reaches the consumer. Rats and mice alone eat or spoil 20 per cent — around two billion tons a year. Inadequate transport systems, insecure warehousing and the lack of temperature controls make up the rest — in hot climates food grains rot in storage before they can be distributed. A major priority could be to subsidize the building of vermin-proof air-conditioned storage for food and air-conditioned transport to distribute it. Such things can be done — a lack of capital is the main reason why they aren't. Maybe someone affluent could provide seed money for this purpose as a loan —this could be reliably repaid from the proceeds from food saved. The urgency of world poverty would be greatly reduced by this one measure alone – at least for the present. Home refrigerators do much to reduce waste – 88 per cent of Chinese homes now have them, only 27 per cent of Indian ones.

Can we produce enough additional food by forcing the land to yield maximum returns by massive use of chemical fertilizers? These – especially nitrates – are polluting rivers and seas and degrading soils to a dangerous extent, with a UN convention in 2013 warning that land three times the size of Switzerland, and valued at almost $500 billion, is being lost every year. Severe degradation almost to a desert condition was reported in 168 countries, yields are falling in many of these places, and violence and conflict is increasing. The most severe problems are in sub-Saharan Africa.

Sustainable means to reduce soil degradation and improve soil that is already damaged are well known and need only money and effort to apply. Many are low technology – planting nitrogen-producing legumes, conservation and controlled use of animal manures, community composting, the restoration of forests and reductions in the use of wood for a fuel by promoting low cost solar cookers, metal discs which concentrate the sun's heat enough to cook food or boil water.

Farmers in the main food-producing countries feel they have their act together pretty well, but there is still room for improvement. Going along with, rather than fighting nature, is now being recognized as sensible. Persistent use of superphosphate makes crops grow better, but in time reduces trace elements, depletes soil micro-organisms, and changes the soil structure, making it drier and more subject to wind erosion and less able to absorb rainwater. It also contains the heavy metal cadmium, which gets into food products.

Organic farming, which sequesters carbon into humus in the soil, can produce fruit and vegetables with a better vitamin and mineral content than chemically grown ones. It also uses much less water, since humus retains water like a sponge. Half of the food for Cuba's capital, Havana, is grown organically on small plots inside the city. Organic fertilizers, derived from animal products, are now coming into greater use. Deep ploughing, once almost universal, is now seen to be bad for the soil in several respects – less invasive tillage to a depth of about eight inches is better, shelter belts of trees should be retained rather than clear felling, and there have been changes to that increasingly vital factor, the use of water. Farmers have traditionally cleared watercourses, removing trees on the banks and allowing streams to run unobstructed. But now it is recognized that slowing currents, actually obstructing the river, allows rainwater to be distributed more evenly through the surrounding soil, and limits erosion. Properties worked this way, restoring streams to their natural condition, have achieved surprising increases in productivity.

'If I didn't do this, who would?' Britain's Prince Charles is said to have observed, when in 1986 he converted 900 acres of his Home Farm to growing organic food. Although he was subject to some disapproval, and even mild derision, 30 years later the project is regarded as a model for sustainable agriculture, and attracts many visitors. Home Farm does not use nitrogenous fertilizers or chemical poisons, but rather aims to build up and keep the natural fertility of the soil. Red clover, which takes nitrogen from the air and fixes it in the soil, is part of a system of crop rotation. Hedgerows are used for fences, power is provided by more than 400 solar panels. Commenting in 2013 the prince asserted: 'Organic can feed the world.' Speaking at a conference on the future of food, he quoted a 2008 UN report involving 400 scientists that found small family farms were the most productive food systems in the world — 'and yet the conclusions of this exhaustive report seem to have vanished without trace.'

Sustainable agriculture basically requires organic matter to be recycled back into the soil, rather than sent to landfill or dumped into the sea, as is now common practice. This can be achieved by composting or growing a cover crop designed to be ploughed in. Modern composting toilets, many thousands of which are in use, can recycle human waste safely and efficiently – large-scale use of traditional water closets depletes nutrients massively by pumping them into the sea. Monoculture, growing the same crop on large areas of land year after year, reduces soil quality – growing many different crops together is preferable. The objective should be to develop farms that are able to 'produce perpetually.' Some of these might be as small as a community garden – even a rooftop garden — in an urban area.

What has been regarded as country too arid to use may be a major new source of food in the future, provided it is near the sea. Many of these soils have productive potential, provided you can bring water to them. This is the case in arid belts bordering the sea in parts of the Middle East, and in South Australia, where Sundrop Farms has built a productive greenhouse powered only by the sun and using desalinated seawater for irrigation. The half-acre facility, which is powered by a solar thermal array, grows capsicums and tomatoes. The solar system desalinates seawater for irrigation and provides heating and cooling. This prototype has been so successful that a 50 acre facility, due for completion in 2016,was being built at a cost of around $110million, to produce 15,000 tons of tomatoes a year for the Australian market and employ 200 people. Similar technology is being developed in Qatar.

The fact that 50 acres of what was considered useless land can produce huge amounts of fresh food is tremendously important, and seems certain to inform the future. And there are other solar technologies that can produce food in desert-like regions. Solar towers, which use a strong draught rising through a high chimney to generate electricity, derive warm air from a surrounding greenhouse, in which food can be grown. Towers like these are planned for Australia, India and the United

States. But sometimes quite low tech methods can prove effective – Stephen Briggs, who farms in eastern England, has increased productivity by planting apple trees in his wheat-field, leading to a much more efficient use of nutrients, since the tree roots feed below the cereal crop. The trees act as windbreaks, and attract pollinating insects.

The fourth and fifth major objectives could be achieved by so-called 'cultured meat', with work going on in several research labs around the world. 'Animal farming is absurdly destructive and completely unsustainable', according to Patrick Brown, the founder of Impossible Foods, a Silicon Valley project that has raised $75 million to develop plant-based cheese and meat imitations. Another, at Maastricht University, demonstrated a lab-grown burger in 2013 at an event in London. This, and the research behind it, were substantially paid for by Google co-founder and billionaire Sergey Brin, and the project was led by Professor Mark Post. Tasters at a public demonstration thought what some of the media described as the 'frankenburger' 'was getting close to meat,' but was slightly unfamiliar and bland, due mainly to an absence of fat. An improved version was being developed in 2015. According to an Oxford University study in 2011 traditional meat production currently accounts for a third of all land use, 18 per cent of global emissions and 27 per cent of water use. Transferring these resources to non-animal food production could eliminate hunger for millions of people.

Meanwhile, a complete substitute for food is already up and running. Called Soylent, from the 1970s SF movie *Soylent Green,* it comes as a powder to which you add water and flavour. American researcher Rob Rhinehart, who partly lived from the product while he developed the process, and who says he has 140,000 orders for it in the US, claimed his health, brainpower, and stamina had improved while using Soylent. Rhinehart, who has turned his kitchen into a library, and has neither a refrigerator nor dishes, says he still eats 'recreational meals' occasionally. A week's supply of Soylent costs $65 – around $3 a 'meal' but this price is expected to drop over time.

12 Water

Millions of icy comets that bombarded the primeval earth have been credited with bringing Earth's water to us from far out in space — but this idea demanded a closer look when the Philae spacecraft landed on comet 67P in 2014. It found water there, but it was significantly different from that on earth — 'heavy water' containing three times as much deuterium. So is it now more likely water came from incoming asteroids, many of which are known to contain water? In 2015 the Japanese space agency was planning landings on an asteroid to settle this matter, among others.

Most of this strange and indispensable substance is in the salty oceans or locked up in ice —the amount available as fresh water is less than three per cent, and is unevenly distributed. The largest reserve, almost 20 per cent, is in Russia's Lake Baikal, the world's largest lake, in places more than a mile deep, the least is in Australia, the driest inhabited continent, where the two largest permanent lakes are manmade.

But perhaps the most unexpected large reserves lie under what is apparently one of the world's driest regions – the Sahara Desert. This huge complex of aquifers, called the Savornin Sea after its discoverer, contains on some estimates about half a million cubic miles of water, sometimes salty, sometimes emerging at temperatures as high as 80C, but mostly useable. Some of it is said to be a million years old. Fertile oases in the desert are upwellings of this vast reservoir, but it is being tapped for much larger flows by several nations, including Egypt, Libya and Algeria. Former Libyan leader Muammar Gaddafi spent as much as $75 billion on his Great Manmade River, which, with almost 2000 miles of piping and aqueducts, he claimed to be the biggest such system in the world. It takes water from the desert aquifers to Libya's major cities and food-growing regions, which are near the Mediterranean coast. In a world plagued by misguided military activity NATO aircraft actually bombed parts of this vital Libyan infrastructure, on which 70 per cent of the population depend for water, as part of the Western intervention in the disastrous Libyan revolution. The subsequent descent of that country into anarchy places this ongoing giant project in jeopardy.

We live on a water world – more than 70 per cent of the surface of Earth is sea. Because there is so much water, the oceans are a major influence, especially the Pacific, occupying almost half the planet. A significant climate element, el Nino, is driven by changing temperatures in the Pacific. Sometimes this influence is weak, but the most extreme events influence the weather in two-thirds of the world, mostly adversely. The last major el Nino, in 1998, released huge amounts of heat into the atmosphere, raising average world temperature permanently by a third of a degree. It also caused weather extremes, droughts, floods and massive bushfires in many countries. At the time of writing, mid

2015, another major el Nino was said to be developing and similar conditions were forecast, according to the US National Oceanic Atmospheric Administration (NOAA).

Year to year variations in air temperature are, then, less significant than the temperature of the oceans, which are an enormous reservoir of heat. On one estimate 90 per cent of the planet's heat gain caused by greenhouse gases goes into the seas, seven per cent to land and ice, and only three per cent to the air – most of the natural variability in surface air temperature from year to year is due to heat moving between the oceans and the atmosphere, rather than any overall gain or loss of heat by the entire planet.

The next proposition is so important to our future it is worth repeating: The oceans, having acquired heat, give it up only very slowly, even in conditions that favour this. 'Very slowly' means spans of many hundreds of years, perhaps as much as a thousand years. So if we allow global warming to continue and perhaps accelerate, we will dangerously affect a dozen generations beyond us. This is in the back of my mind as I write every word of this book — that we have it within our power to grapple with the Steamroller, but if we squib it we condemn our descendants to lives more difficult and painful than anything we can imagine.

According to the UN, as much as two-thirds of the world's population will be afflicted with water problems by 2025 as rainfall patterns are distorted by global warming. Already millions are not getting enough clean water to drink, but there will be other consequences. The effects on agriculture and food production are obvious, but many businesses need lots of water to keep going. In 2013 Canada's Barrick, one of the world's biggest gold miners, suspended Pascua Lama, a huge mine high up in the Andes estimated to produce nearly a million ounces of gold in its first five years, because of local concerns about downstream water contamination and possible impacts on nearby glaciers. Mothballing the project has already cost $5billion. Shale gas extraction and coal-fired and nuclear power stations are already coming into question in many parts of the world because they use a lot of water, which generally needs to be shared with urban and agricultural consumers.

Since negative consequences of global warming are almost certain to extend for centuries in an overcrowded and heavily-armed world, future water wars are only too likely. An Israeli minister once told me that water supply was one of the most compelling worries of his government. 'Future wars' he said, 'will be about water – they may not seem to be, but they will.' These words gained force decades later when Middle East security analysts said in 2014 that the outcome of the wars in Syria and Iraq might rest on who controlled the region's already scarce water supplies – this could be an even more important factor than who controlled the oil refineries. At that time daesh, the Islamic organization claiming to launch a new 'caliphate', controlled the headwaters of the two great rivers of the region, the Euphrates and the Tigris.

The level of the Sea of Galilee (Kinneret) is monitored with great anxiety, since it is the major natural source of Israel's water. It has not filled to flood levels since 1992, and in 2015 was approaching crisis level at almost 700 feet below sea level. The Jordan River, of Biblical fame, which flows into and from the Kinneret, is little more than a small, polluted drain. I can vividly recall standing on the Allenby Bridge, not far from where the Jordan enters the Dead Sea, looking down with surprise and disgust at the stinking trickle of water below. The intensely saline lake it feeds – the Dead Sea – is surrounded by blinding flats of white salt as the water surface steadily reduces.

Israel recycles 86 per cent of its domestic wastewater, and is building huge desalination plants to guarantee future delivery of the 525 billion gallons of water consumed a year – half of it for agriculture. In 2010 130 billion gallons were desalinated, with plans to increase this to 200 billion by 2020. The more than 10,000 desalination plants in 130 countries are hungry consumers of electric power, mostly generated using fossil fuels. This is plainly unsustainable, so solar power is now being developed for this purpose. A number of smaller units around the world, like some in the Greek islands, are using photovoltaic power. Desalination plants in the Australian cities of Perth and Sydney are fuelled by wind-farms.

The growing demand for water and its uneven distribution are making sustainable power for desalination an urgent requirement, especially in the

Middle East, which is one of the most water-hungry regions on earth. Egypt wants to quadruple its productive land by irrigating parts of its western desert with a system of canals and pumping stations. However, the water for this New Valley project has to come from the Nile, which already has just enough water for the ten nations on the river. Some of these, as far away as Kenya, are already protesting. After 30 years this $90 billion project had made little progress by 2015, encountering severe problems, such as the high salinity of desert land proposed for irrigation, and its future seemed in doubt.

China, with acute problems supplying water to at least 100 of its cities, is working on the world's largest engineering project, a vast canal and aqueduct system more than 1000 miles long to bring water from the south to the parched north of the country – a waterway twice the length of the Grand Canal in its heyday. This will cost over $60 billion dollars. Meanwhile the government is struggling to control severe water contamination of 200 rivers, many of which provide water for large cities.

Water shortage may well turn out to be China's worst problem —according to a former water minister, Wang Shuchen, quoted in *The Economist* in 2013: 'We have to fight for every drop of water or die; that is the challenge facing China.' Millions of Chinese have barely a third of what the average American uses. In the capital, Beijing, the water table has dropped nearly 1000 feet in the last 40 years. Around half of all Chinese live in the north and two-thirds of the farmland is located there, but

80 per cent of the country's water is in the south. Northern groundwater is in a deplorable state, with 70 per cent so polluted humans cannot safely contact it, even for washing, much less drink it. A third of the water in the Yellow River is unfit even for agriculture, but is so over-used there are frequent days when water never reaches the sea.

Water tables are falling or drying up in so many places this was assessed in 2013 by Lester Brown, of the Earth Policy Institute in Washington, as a major threat to global food supplies. He said the most serious problems are in the Middle East, with declining water supplies in Saudi Arabia, Syria, Iraq and Yemen, and claimed that 18 countries containing half the world's people are over-pumping their underground water tables to the point where they are not replenishing and where harvests are getting smaller every year. These aquifers only be replenish very slowly, and all over the world water   is being taken from them faster than it can be replaced. As the vast Ogallala aquifer in the great plains of the United States shrinks, thousands of square miles of farmland face the prospect of reverting to desert – in the words of a 2015 Bloomberg headline 'a looming water crisis.' In this matter, as in others, time is against us. As the great aquifers become exhausted, world food production must drop significantly.

India has heavy monsoonal rain, but experiences regular water shortages because of a lack of adequate storage facilities. Women in country areas typically walk almost 1000 miles a year just to get water, which they often have to carry several miles every day. The source, poorly-maintained wells, frequently yields nothing better than dirty, saline and contaminated water. The majority of India's farmers are smallholders on perhaps four acres, with only precarious access to rapidly diminishing supplies of groundwater for irrigation, which for most is their only source of water. They are having to drive tube-wells up to 800feet below ground to reach falling water tables – cheap or at times free electricity allows them to do this – but even so in many places groundwater reserves will dry up altogether in five or six years. In Tamil Nadu, in India's south, with 62 million people, wells are drying up rapidly, reducing the state's irrigated area by a half in a decade. Even the capital, Delhi, is facing a permanent water crisis, with the water table dropping three feet a year. Wells in neighbouring Pakistan are three to six feet lower every year.

Recent research has confirmed that most of the 1500 Himalayan glaciers that feed Asia's big rivers are shrinking. Satellite observation by the European Space Agency in 2007 and 2012 showed retreat of from 30 to 50 feet a year, the fastest glacier regression in the world. Observers believe two-thirds of the glaciers may disappear by 2050, and all may be gone by 2100. The ESA is continuing these observations on a regular basis, recognizing their enormous significance in assessing the world's food situation. This is because a regular flow from these glaciers, the biggest water resource outside the polar icecaps, supports billions of people in south and east Asia. The glaciers 'buffer' rain and snow falls in the headwaters of seven major rivers, including the Ganges, the Yangzi and the Irrawaddy. According to one expert, Professor B.G Verghese, of the Centre for Policy Research, New Delhi, there will be a greater flow of water from glacial melt for about 30 years, followed by a 30 per cent drop – 'that is really going to hit us.' However, at least the Chinese are doing something about it— they plan to build 59 reservoirs below major glaciers in the Tian, Kunlun and Altai mountains to collect melt-water.

China, India, Pakistan, Bhutan and Nepal plan to build 400 hydroelectric dams on the rivers flowing from the Tibetan plateau – many of these will be close to the sources of Asia's great rivers. This immense building programme, designed to produce almost a quarter of a million megawatts of power, is an important part of the world effort to produce electricity from non-polluting sources, but its impact on river flows and fisheries on which 40 per cent of the world's people rely is unpredictable.

The concept of 'virtual water' is one of those things that seems obvious once someone – in this case Professor J.A. Allan, of King's College London, has thought about it. Roughly a sixth of all the water used in the world grows plants and livestock which are exported in one form or another – so it's not only water that can make you wet that is important — if you export oranges, you are losing the perhaps scarce water it takes to grow them. This fact was put forward by an Israeli agricultural economist, Gideon Fishelson of Tel Aviv University, and later taken up by Allan, who realised the argument was two-sided — the exporting country gave away its water, but the importing country conserved its own supply by importing most of its food. The classic case is Singapore, which has severely restricted natural resources of water, but which has developed so much wealth through its high tech economy it can afford to buy in almost all its food.

The other biggest net importers of virtual water are China, Japan, much of Europe, Egypt and South Korea. The largest exporters are the producers of food grains and meat, like the United States, Argentina, Canada and Australia. All this is important because virtual water goes a long way towards evening out the world's supply problems, taking water from places that have plenty to others with problems. This has been the case in the past, but as supply tightens, will it make sense to export huge quantities of wheat and beef? It takes 20,000 litres of water to produce one kilo of beef, about 1000 for wheat, more than 2000 for rice – and 140 are needed to produce one cup of coffee.

Contaminated water is deadly – every year it kills more than two million people from waterborne diseases. Water becomes toxic when animal or human faeces, carrying a depressingly long list of nasty organisms, enter it. One such A to Z list contains over 30 items. The World Health Organization says more than two billion people are infected with something you may never have heard of – schistosomes. These are flukes —tiny worms

— that live in the bloodstream of infected people, mostly children. These parasites, which are carried by freshwater snails in natural bodies of water, can infect animals and humans by direct penetration through the skin. Liver fluke is a similar common infection, which results from eating raw or under-cooked fish, causing fatigue and stomach discomfort. If left untreated for years, it can destroy the liver. Then there are the 'tropical' fevers like cholera and typhoid, which are frequently carried in water.

But diarrhoea is the big killer of under-fives, causing at least half a million child deaths a year, around ten per cent of all infant fatalities. The infected water these children are given to drink is usually the only water their parents can get, so there is a dreadful inevitability about these horrifying figures — this could only be remedied by financing better sanitation and clean water availability.

However, disease germs are not the only dangerous ingredient of water. After Bangladesh became independent massive aid programmes were started to deal with almost universal waterborne disease. More than three million deep tube wells were bored to access what seemed to be safe artesian water. However, it was soon found most of this water contained dangerous quantities of arsenic, which causes skin complaints, cancer and eventually death. This has been described as the largest poisoning of a population in history, with as many as 70million people affected. The problem is not unique to Bangladesh —possibly 150 million people around the world may be affected, with almost 20million Chinese living in high-risk areas. Filtration and chemical treatment of the water are being introduced to cope with this problem, but are still far from universal.

Finally, let's turn to water in its solid form, ice. Until quite recently its main interest was to put in your drink, but now ice has become very important because the amount of it and its location are changing. For 10,000 years from 20,00BCE the weather veered between extremes, a kind of savage seesaw during which, on some estimates, our numbers dropped to a mere 4000. So we have been lucky – or opportunistic – because human civilization has prospered over the past 12,000 years in untypical conditions. These will not last. Some very complex calculations have been made to predict when the next ice age is due, but the results vary widely. However, it seems unlikely over the next 10,000 years, with outside estimates as distant as 30,000 years.

13 Trashing the Planet

The farmer and his wife go on a three-month cruise, using an automated system to keep his cabbages watered. However, a continuing problem with snails accelerates in his absence – they breed in such numbers they totally destroy the cabbage plot. Deprived of easy food, they then die of starvation in hundreds. When the farmer gets back there are many fewer snails than when he left — and no cabbages. Substitute the world for the cabbage paddock and humans for snails and you get a picture roughly similar to our current reality. Almost every natural system is under threat now from humans — we seem hell-bent on damaging what sustains us. In 2014 the Zoological Society of London and the World Wildlife Fund reported that the number of wild animals, land and marine, on the earth had halved over the last 40 years. These missing creatures were almost all killed by humans for food.

We have already ruined some of the world's most remarkable ecosystems — Madagascar, a large island off the coast of Africa, once had beautiful forests inhabited by an extraordinary diversity of animals and a prolific and varied flora. Seventy per cent of its forests were destroyed while it was a French colony between 1895 and 1925 when the land was cleared to grow coffee. More than half of the remainder has gone since. Because of this mass deforestation the ecology has been profoundly disturbed, so Madagascar now has difficulty providing water, food and sanitation for its rapidly growing population. Much of the loss has been due to Madagascar's endowment of rare and expensive timbers, like ebony and rosewood, which have been heavily and often illegally logged, sometimes with the connivance of corrupt governments. The island, which is the fourth largest in the world, had only a little over two million people in 1900 — this had grown to 23 million by 2014. Social indicators are poor – there are around four hospital beds per 10,000 people, and an under-five death rate of 65 per thousand.

More than 80 per cent of Madagascar's species did not exist elsewhere in the world – the island is one of the planet's most remarkable examples of biodiversity. Now many have gone or are endangered in an environment of degraded scrubland and near deserts, with rivers running blood red from massive erosion of productive soil. Huge fires, originally started to clear land, burn for weeks as they spread into what is left of the virgin forest. Madagascar is one of the world's 35 'biodiversity hotspots', areas of outstanding natural value in which most of the endangered land vertebrates live – many of the rest of these are faring little better. Among them are the Andes, Borneo, Java and Sumatra. Some have only five per cent of their natural vegetation left, with about 15 per cent remaining within the whole of these 35 hotspots. Their total destruction would mean the loss of half of all species on earth.

I have beside me an article by Ove Hough-Guldberg, Professor of Marine Science at the University of Queensland, in which he predicts that the world's coral reefs 'will become ecologically extinct by the middle to late part of the present century.' His main concern is Australia's Great Barrier Reef, 'a magical realm, endless in geography and exquisite in time and space ... something special to treasure and preserve for all time.' The world's largest complex of coral reefs, it stretches for more than a thousand miles, supporting a tourist industry worth billions of dollars — yet in spite of all this, almost half has been damaged by global warming, and the rest is endangered by the projected development of large ports to export coal to India and China.

In the inland of Queensland the Galilee Basin is one of the largest undeveloped resources of coal in the world – it contains literally billions of tons. Nine large mines proposed for the region are owned mostly by multi-national corporations, who are building railway lines to the coast, opposite the Great Barrier Reef, where Abbot Point will become one of the largest coal-loading ports in the world. It is only one of a series of megaports planned along the coast. Thousands of bulk carrying ships will move through sensitive areas to export a staggering 330 million tons of coal a year – environmentalists fear accidents will be inevitable, with attendant oil and coal dust pollution and physical damage to the fragile reef.

Coral, a tiny organism, lives in a symbiotic relationship with an algae that provides it with the amino acids and sugars it needs to survive and to deposit the huge amounts of calcium carbonate –lime – that make up the reefs. Without this activity new reef frameworks cannot develop and existing structures fail to repair the inevitable damage caused by wave erosion and predators. The algae are dying because the oceans are getting warmer. They also provide the reef with its colour, so once they are gone the coral becomes bleached — a dirty and drab greyish whiteness. This phenomenon was first noticed in the Caribbean in the nineteen-eighties. In Hoegh-Guldberg's words 'by the end of 1998, an estimated 16 per cent of corals had been extinguished from the world's reefs, with areas that experienced the most extreme temperature having mortality of up to 93 per cent of all corals within a region.'

He says sea temperatures by the middle of this century, even under relatively mild carbon dioxide emission scenarios, are likely to trigger serious mass coral bleaching and mortality. This means coral-dominated reefs are likely to be extinct by 2050. Many corals are also sensitive to increasing acidification of the oceans. This higher concentration of carbonic acid, derived from CO_2, makes it more difficult for critically important sea life to grow shells and skeletons. Already research has shown that these natural systems are severely disturbed. It matters that creatures so small you can hardly see them can't grow a shell, or are adversely affected by higher water temperatures. They are at the bottom of the food chain — if their numbers are reduced all marine life higher in the chain must also be affected – and, eventually, us.

These things are bad enough, but there is a more basic long-term risk if we persist in business and life styles that increase the level of greenhouse gases in the air beyond a certain point. People who have serious asthma and those around them are only too aware of the terrifying incidents caused by a failure of the lungs to process oxygen from the air. This can be life threatening, unless it is corrected by the provision of oxygen from an outside source. Like most life on earth, we are dependent on this highly corrosive and in many respects dangerous element. Look at the body of a car severely affected by rust and you'll understand how corrosive oxygen is. Although it can be just as destructive to our bodies, we are nevertheless adapted to it, to an extent that without a certain amount of it in the air we sicken, lose consciousness, and eventually die.

In pre-human times oxygen probably made up as much as 35 per cent of the air – now it is 21 per cent. If that percentage falls below 18 per cent we don't feel well. Let oxygen fall to 7 per cent and we are dead — not only us but almost all other living things around us. Oxygen is not a stable lasting commodity, constantly recycled like water – natural events can reduce it, just as they can make it. Plants absorb carbon dioxide and use it to manufacture their components, like wood and all their products that we use as food. They give off oxygen – so there is a symbiotic relationship between them and us that normally preserves the balance of these two gases in the air. Microscopic plant life in the sea is a part of this process. Phytoplanktons, small sea creatures that have the power of photosynthesis, make at least half our oxygen.

People admire tropical seawater that is blue and transparent, but this really means it is barren and relatively devoid of life. It is opaque grey-green water that is productive, like that of the huge cod fisheries of the north Atlantic fisheries. It gets its colour from dense concentrations of plankton – the bottom of the marine food chain. Climate change is reducing phytoplankton numbers, slowly but inexorably – the greatest loss is in polar and tropical waters. However, you are not going to choke to death soon — according to the Woods Hole Observatory in the United States planetary oxygen levels fell only 0.0317 per cent between 1990 and 2008 — there seems to be some natural compensation factor in play that is not fully understood. According to Professor Ralph Keeling, who heads the Scripps Institution's o2 Program 'oxygen levels are decreasing globally due to fossil fuel burning ... but the changes are too small to have an impact on human health.' Also humans can, over time, adapt, as do four million Tibetans who live at thin air altitudes above 10,000 feet. Even so, there are places inhabited by millions of people where oxygen levels are already dangerously low – down to 15 per cent – and this is probably contributing to high rates of fatal disease. The worst are the world's biggest cities, like Delhi and Beijing, where oxygen depletion is associated with the heavy load of air pollution now so typical. The people who live in these cities are almost certainly experiencing early symptoms of oxygen deprivation, *hypoxia,* which, in the words of an experienced flight surgeon, means stupidity – an aircraft pilot exposed to low oxygen gets too dangerously stupid to see when something is wrong.

As with so many other things, the eventual deciding factor for oxygen levels is going to be the *extent* to which our activities are damaging, and that, of course, is something we should be able to control. Consider this set of numbers, which are arguably the most important you will ever read: 2,3, 3.5, 4, 4.5, 5, 6. These represent the range of increases in the planet's overall temperature over pre-industrial levels, measured in degrees Centigrade, that climatologists believe are possible this century. Generally speaking the world could tolerate a 2 degree increase, perhaps even 3 degrees, although this would involve a great deal of danger and discomfort, including much wilder weather than we have ever experienced. Beyond that is a matter of speculation because we would be encountering extremes of heat, especially in the interiors of the continents, at levels previously unknown to us.

The figures set out above represent an *average,* while of course temperatures fluctuate greatly in different parts of the world. Land areas get a lot hotter than the sea. Hence an average 4 degree rise could result in as much as 10 degrees in many places now inhabited and used by man. An inland city that has tolerated say, 39C, for short spaces of time, could be struck by close on 50 degrees. Prolonged periods of anti-cyclonic weather, with clear skies and little wind, in places far from the coast can produce abnormal heat for extended periods.

Dry winds on the lee side of mountain ranges can do the same. Cities, with large areas of black roads, parking areas, and often dark-coloured roofs, can become 'urban heat islands,' hotter than the surrounding countryside by a possibly critical 5C. Planting trees, and painting roofs and roads in light colours can reduce this effect.

Our normal body core temperature is around 37.5C, but problems arise if it goes much above 40. The body can get rid of heat by sweating, including evaporation of moisture in the lungs, so outside temperatures of perhaps 44C can be tolerated by a young, healthy person, but not much more than that. So we would do well to regard temperature readings approaching 50, with moderate to high humidity, as dangerous. These can kill older people in large numbers, especially if they have chronic illnesses. The European heat wave of 2003, which affected 12 countries, caused 70,000 deaths, on a number of reliable estimates. The Russian heat wave of 2010 killed 55,000. Many of these were 'premature' deaths –that is, of old and ill people. In places with a cooler climate, like Britain, anything over 30C is regarded as a problem.

Donana National Park in Spain, 200 square miles of saltmarsh, shallow creeks and sand dunes, is one of Europe's most important wetlands, a World Heritage site that is home to a huge variety of birdlife. As many as 200,000 migratory waterfowl have been observed there in winter, 300 different bird species use it. It attracts nearly half a million visitors every year. However, when a tailings dam at the giant Los Frailes copper, lead and silver mine upriver from the park failed in 1998, huge volumes of contaminated water flooded into Donana, causing Spain's worst environmental disaster – it ultimately cost over $100 million to repair. The highly acidic water, containing arsenic and cadmium, killed thousands of birds and fish, and marred what had been a pristine and beautiful landscape. The mine was closed, but because of very high levels of unemployment in Spain in 2014 a decision was taken to reopen it. It was the last of a series of economic decisions, all of which will again impact Donana – a kind of environmental slow burn. An oil pipeline may be built through it, water-hungry strawberry farms are drying it out, silt from other farms is blocking its channels and new hotels, golf courses in the region will place further demands on the water table. Urban development at the nearby port of Huelva is considered a threat.

London-based company SOCO International wants to drill for oil in the Virunga National Park, one of the largest and most important reserves in Africa, covering almost two million square miles. In 2015 the park's chief warden, Emmanuele de Merode, was recovering in hospital after being shot in an ambush, while the World Wildlife Fund reported two of its employees had been threatened in anonymous phone calls because of their opposition to oil drilling. According to UNESCO's Guy Debonnet Virunga is an important test case because it is the first of the world's 193 natural World Heritage sites to be licensed for drilling. Both UNESCO and the World Heritage

Committee have asked the government of the Democrat Republic of Congo to set aside oil exploration permits they have granted in the park.

The predicament of these places is typical of a struggle going on all over the world, as worsening financial conditions are forcing economic development at the cost of the environment. One of the most intense of these concerns palm oil, which is used in cosmetics, ice cream, margarine and huge quantities of processed foods –almost half of all supermarket products contain it. In Malaysia and Indonesia, and more recently in Africa, the economic consideration has come out on top, as more and more land is being cleared to plant oil palms, serving an industry whose global production is valued at more than $50 billion. There is a considerable debate going on about the environmental hazards of this trade, including concern that the modern world could share the fate of ancient civilizations known to have collapsed because they over-exploited their resources.

Extinction of species is an ugly phrase, but it's the right one. Five times in the past there have been massive wipeouts of species, all due to natural causes. The last of them, 65 million years ago, destroyed three-quarters of the creatures then on the earth, including the dinosaurs, and was probably caused by a meteor strike. It also provided the space and the opportunity for primitive mammals, then eventually us, to evolve. Now, according to biologist Stuart Pimm of Duke University in the United States, 'We are on the verge of the sixth extinction.' Pimm and his colleagues estimated that before there were human beings on the earth, less than one species per million became extinct every year. Now he says that rate is hugely advanced, to as much as 100, or even 1000 per million. The main reason is that there are too many humans, who take over and adapt habitats so much that creatures living there are eaten or simply die out –'death by a thousand cuts', according to Professor Will Steffen, of the Stockholm Resilience Centre. According to Steffen 'humans are eating away at our own life support systems' at an unprecedented rate – primary energy use has increased fivefold since 1950, fertilizer use eight times, and the amount of nitrogen entering the oceans four times. These figures come from research over five years by two international teams, who found that adverse changes over the last 60 years had had no precedent in the past 10,000 years.

Some studies have concluded that almost 90 per cent of species decline over more than 100 countries is associated with more crowded human populations, which is also not good for the humans. Habitat loss is closely associated with soil degradation, nutrient depletion and pollution. Acid rain, algal blooms and fish kill are among the consequences. And it seems likely to get worse. With estimates that agricultural output will need to grow by at least 50 per cent to meet population growth over the next few decades habitat destruction seems likely to increase, and more and more species will die out because of this.

Once the world had around 4 billion acres of tropical rainforest. That is now down to less than 2 billion, with about 30 million acres being destroyed every year. Almost half of what is left is in South America, mostly in the vast Amazon region. More than 20 per cent of those forests have been cut down for logging and to make room for agriculture over the last half century, and as much again is under threat. This could have disastrous implications for the world's climate, since the Amazon rainforest is one of the world's largest reservoirs of carbon dioxide. Drought and wildfires could reverse that situation quite abruptly, as they did during a long drought in 2005 when the Amazon released almost a billion tons of CO2 into the atmosphere.

Temperate region forests have fared even worse – in the most 'civilized' places barely two to three per cent still survives. More than half of the wetland habitats in the US and Europe have vanished. Twenty per cent of all coral reefs and around 40 per cent of mangrove ecosystems have gone. It's difficult to know how many species are being lost simply because we don't yet know how many there are. According to the World Wildlife Fund there could be 100 million different species on earth – 97 per cent are invertebrates such as earthworms. Of the 17,000 species currently under threat the ones closest to us are most severely at risk – about 20 per cent of primates are fast disappearing, with many facing extinction. Some, like the gorillas, chimpanzees and 'the men of the forest' – that is what *orang utan* means — could be extinct within a human generation. The great apes are of course primates like us — the chimpanzees and bonobos share 99 per cent of our DNA. While loss of habitat is a major factor, hunting for food is another – these animals, in the word of conservationist Russell Mittermeier, 'are quite literally being eaten into extinction.'

This applies particularly to the gentle, reclusive bonobos – mountain gorillas – closer to us than any other species and, in the opinion of many observers, socially superior to us in some respects. Once estimated to have numbered around 200,000, only 50,000 survive now. With long legs, head hair that forms into a part, distinctive differences in face shape and expression more typical of humans, and a co-operative and peaceful social life in which differences are resolved by sexual contact, some anthropologists compare bonobos with early forms of humans, such as *australopithecus*. Bonobos appear to converse verbally. They can learn to understand human conversation and two, Kanzi and Panbanisha, communicate via special keyboards. Scientists at the Great Ape Trust estimate that Kanzi has a vocabulary of more than 500 English words, and can understand around 3000. Like humans, bonobos laugh.

The many thousands of people working to avert extinction of the great apes are also specially concerned at the falling numbers of the tree-dwelling *orang utans* – incidentally that name is pronounced orarng ootarn, the hard sounds commonly used by Westerners don't exist in Indonesian. These animals are also among the most intelligent primates, with an elementary tool-using capacity and word-learning ability. They are an endangered species in their Borneo and Sumatra habitats due to hunting, forest fires and the massive conversion of forest into palm oil plantations.

Most people don't like sharks. We have a visceral fear of being eaten by one – certainly I did when I found myself swimming with sharks off a reef in the northern islands of Fiji. But the locals laughed at me... 'No worry... plenty fish...' What had fish got to do with it, I asked. It seemed no one there had ever been attacked by sharks because they really prefer to eat fish. Yes, they'd heard of shark attacks on people in other places. This was because there weren't enough fish and the sharks were hungry. But even then, after one bite the sharks tended to swim away... 'not like taste of people.' I was far from being convinced by this, but it is a fact that many fewer sharks kill people than people kill sharks – 100million of them every year, according to Sea Shepherd. In many places sharks' teeth are valued for jewellery, their skins are made into belts and wallets, their liver oil goes into cosmetics. But worst of all is taking shark fins for what I consider a very overrated Chinese soup. Many are killed just for the sake of it by 'sport' fishermen, but the 'finning' is by far the nastiest. The fins are cut off the living creature and it is thrown back into the sea to endure a long, agonizing death. Because of these things sharks are an endangered species. They are being killed faster than they can reproduce – the populations of large sharks have been reduced by as much as 90 per cent. These are predators who are vital to the marine ecosystem, as they have been for 400 million years. According to Scientific American (21.1.14) 30 per cent of sharks, rays and associated species are at risk of extinction. Even the huge, much-feared Great White is decreasing in numbers.

14 The Sea

Dead zones, vast rubbish-strewn gyres, green tides, red tides, jellyfish infestation, and toxic plastic in your fish dinner – these are some of the bad things degrading our oceans. At the most basic level our bodies are intimately connected with them — life on the land, including, eventually, us, came from there, and even now we carry its heritage — the mix of minerals in our blood is much the same as in seawater. The oceans cover more than two-thirds of the earth's surface, their vast abyssal plains three miles down comprise almost half of it. These deeps are still places of mystery, largely unknown, the source of legends like the Kraken – a huge octopus, believed to be big enough to swallow ships. In 2011 American palaeontologist Professor Mark McMenahin deduced from fossil finds that something like the Kraken might have existed in the distant past, 'an ancient, very large sort of octopus.'

It is only quite recently that exploration of the deeps has become possible, so now we know strange, unfamiliar life-forms, some brightly coloured, some able to generate electricity to self illuminate, some giant squids of Kraken-like appearance, others with myriad tentacles like the basket star, exist in abyssal water. The glass squid is completely transparent, one deep sea siphonophore has red lamps that light up to attract prey, there are creatures down there of such strange appearance they might be denizens of another planet.

Then there are the archeons, the ancient ones, one of the earliest forms of life, thriving in volcanic water above boiling point, in five hundred atmospheres of pressure.

However, it is the five per cent of the oceans on the coastal shelves that are of the most economic importance to us. As much as a fifth of sea life exists in these shallow seas, 90 per cent of our fisheries are located there, also vast quantities of the tiny marine plants known as phytoplankton, which use photosynthesis, just as land plants do, to provide much of our oxygen. So the welfare of these coastal waters is closely involved with our welfare and sustenance.

Yet in spite of this wonder and deep significance, we continue to contaminate the sea and destroy its creatures at a breakneck speed. Rubbish, trash, garbage, call it your way, fouls our oceans to an extraordinary extent – video and trawl samples taken in 2014 at 32 sites in the Atlantic, Arctic and Mediterranean found bottles, plastic bags and fishing nets on the sea floor up to three miles below the ocean surface. Almost half this rubbish is plastic and around a third fishing debris – the Five Gyres Project estimated the number of plastic bits in the oceans in 2014 at over a trillion.

The Great Pacific garbage patch sits within an area of circular currents called a gyre, and occupies a space that amounts to at least a quarter of a million square miles. Other major rubbish patches in the Indian Ocean and the Atlantic are estimated to be hundreds of miles across, with half a million bits of debris in each square mile — again, much of this rubbish is plastics. Some of these degrade within a year, and can deposit toxic chemicals like PCBs and bisphenol A into the sea. Eventually some common plastics disintegrate into tiny particles, which are then ingested by fish. So far eight species have been identified containing toxic particles. In 2014 research in Hawaii found that more than half the catch of a commercial fish, *opah,* (moonfish) people were eating had ingested plastic. According to seven researchers who reported to *Environmental Science and Technology* in 2011 many of these plastic fragments – as many as 1900 from a single garment —come from fibres rinsed off in domestic washing machines during wash cycles, then transferred to the sea in wastewater. At the time of writing research was beginning on whether toxic chemicals in the fish were being transferred to people – there is already evidence that they cause liver damage in fish.

On some estimates over 7 million tons of plastic are circulating in the five main oceanic gyres, although Greenpeace estimates ten million tons enter the sea every year. The most celebrated ocean rubbish patch is probably the Sargasso Sea in the North Atlantic, a region off the American coast, where gyre currents have accumulated masses of seaweed famous since the time of Columbus. In spite of the myths, this weed does not trap ships – the idea of 'a graveyard of ships' is purely fictional. However it too has accumulated vast amounts of plastic and other rubbish. Marcus Eriksen and Anna Cummins of the Five Gyres Project found plenty when they sailed through it, toothbrushes, bottle tops, cigarette lighters as well as millions of confetti-sized fragments.

Gyres are very large, slow-moving whirlpools –
but of course there are faster ones, caused by the meeting
of opposing currents, that can reach speeds over 20 miles
per hour. Again, despite the myths, these are never
strong enough to sink ships. Freak, or rogue, waves are
much more dangerous, and are credited with the
otherwise unexplained loss of dozens of ships. These are
waves much larger than others in the same environment,
and seem to be caused by several waves merging into
one much larger one. Children have been plucked from
beaches and drowned by these monsters, which in mid-
ocean are credited with reaching 100 feet in height. One
of the world's biggest liners, the *Queen Mary,* came
close to foundering when she was hit broadside on by a
90 foot wave in 1942, and the 90,000 ton freighter
Derbyshire disappeared without trace in 1980, the most
likely explanation being that she encountered waves
approaching 100 feet.

For tens of thousands of years, our far-off
ancestors clung to the shorelines because a major part of
their food and commerce was fish. Perhaps, like me, you
thought the major source of protein food was beef, or
maybe pork or mutton. Actually it has been fish since
prehistoric times, and it's still fish, a staple for the three
billion or so humans who live within a hundred miles of
the sea. Most of these people are too poor to afford
animal proteins, so they are dependent for this vital
nutrient on seafood, which they catch off their coastlines.
Or used to. Instead the fish are being taken in huge

quantities by invaders from far away — large ships owned by the 'developed' nations, whose 'industrial fishing' uses techniques that can make virtually a clean sweep of anything in the sea, using drift nets up to 35 miles long hanging down from the sea surface. In 1991 the UN banned all drift nets longer than a mile and a half, but this is still poorly policed.

Large predatory fish like shark, tuna and swordfish are especially vulnerable to over-fishing. Their numbers have dropped disastrously – as much as 90 per cent over the last 50 years on some estimates. The worst damage is done on the high seas, those waters outside the 200 mile territorial regions off coastlines. One-third of fish stocks in national waters are considered to be over-exploited, but this gets closer to two-thirds in international waters, where as much as a quarter of the world catch is illegal, for want of any effective control. There is no international register of fishing boats and, according to the UN's Food and Agriculture Organization, around half of the world's nations admit to having no supervision over fishing vessels sailing under their flags.

And 'industrial fishing' is heavily subsidized. These big ships travelling large distances from their home countries use a lot of diesel – subsidizing this and other concessions amounts to more than $30 billion a year, approaching half of the total value of the catch. Without this massive assistance from rich world governments, 'industrial fishing' would be a heavy loser financially. Corrupt governments in many of the poorer nations of Africa, Asia and Latin America are paid off to permit 'industrial fishing' off their coasts, in spite of protests from local fishermen. The only effective answer is for governments to keep out destructive influences by putting up legal barriers. Marine scientists want fishing prohibited in a significant part of the oceans – this would allow fish populations inside these zones to recover and as much as double their numbers in seven years.

On the coast of Senegal not far south of the capital, Dakar, thousands of women and children have traditionally gathered every day to sort, dry and salt fish caught from hundreds of *pirogues,* small fishing canoes – this traditionally has been the major industry in the region. Much of this salt fish is sold in markets inland in Africa, as much as a thousand miles away. But during the 1990s large foreign ships began arriving, some licensed, some illegal, and fish stocks began to dwindle. Then more recently a dozen foreign-owned factories began converting fish into meal, mostly as feedstock for fish farms in Europe and Asia. One of the most recent of these processing factories, which is Russian-owned, would need almost 500 tons of fish to reach full production, although fish landed by local boats has fallen to barely 200 tons a day.

Disruption of traditional local fishing is happening in many other parts of the world. In the beautiful 'coral triangle' in the Pacific – more than 3 million square miles of reef and island-studded seas centred on Indonesia – stocks of reef fish are down so far that local fishermen's families often go hungry, yet seafood exports from Indonesia to America, Europe and Asia were worth almost $4billion in 2012. Over-fishing is only part of the problem. Huge areas of forest have been cut down in Indonesia and replaced by palm oil plantations. This has resulted in nutrient run-offs that are destroying the region's coral reefs, already vulnerable to bleaching and ocean acidification. While the nations of the region decided in 2007 to set up marine protected areas, most of these still existed only on paper a decade later, or were ineffective because of inadequate policing.

Over-fishing of North Sea herring brought stocks almost to extinction in 1975, forcing a shutdown of the fishery in the hope that breeding could provide a natural recovery. The cod fishery, producing 300,000 tons a year in the 1980s, had fallen to 100,000 tons in 1999, and in spite of reduced quotas, has never really recovered. These smaller catches have been an economic disaster for villages and towns that have relied on fishing as a main industry for centuries. Fifty years ago there were 50,000 fishermen in Britain — now there are about 15,000. Over-fishing of the Atlantic cod cost the jobs of 30,000 Canadian fishermen after the Grand Banks fishery was shut down in 1992 when cod numbers fell to 1 per cent of the traditional population. Cod numbers are said to be recovering but are still no better than 10 per cent of the original stock.

Human activity on land contributes massively to ocean contamination — over-use of nitrogenous and phosphorous fertilizers results in a run-off of excess nutrients into the sea. These favour rapid growth of algae, especially the notorious so-called 'blue-green algae' (actually a bacterium, cyanobacteria) and reduce oxygen levels in the water enough to destroy or discourage all sea life except jellyfish. This chemical overload, combined with the warming of the oceans, results in the notorious biological deserts known as 'dead zones', of which there are now more than 400 around the world. The largest, like one in the Gulf of Mexico, cover thousands of square miles. The Mississippi River, which drains almost half of the continental United States, dumps huge loads of excess fertilizer into the gulf. Dead zones are on the increase — there were 120 in the 1980s, 405 in 2014. They kill fish in the billions. The Baltic Sea has seven of the ten largest dead zones on the planet and beaches on the Yellow Sea coast near the Chinese city of Qingdao are regularly covered in layers of the dark green slime so typical of cyanobacteria. Inland waters are not exempt; the American Lake Erie had a blue-green algae infestation in 2011 that covered almost 20 per cent of its surface.

Seagrass meadows are being destroyed all over the world by coastal development and nutrient runoff, and until now not much notice has been taken of this. However, Australian University of Technology scientist Dr Peter Macreadie's research indicates that seagrass is 35 times better at capturing and storing CO_2 than rainforest.

Because jellyfish can tolerate much lower oxygen levels in water than other marine life, their numbers are increasing alarmingly. Some huge ones, several feet across, have blocked the water intakes of nuclear power houses and disrupted the fishing industry in Japan. Others are stingers. Every year 150,000 people are treated for stings on the tourist beaches of the Mediterranean. This landlocked sea is virtually tideless, with little self-cleaning capacity, and remains heavily polluted in spite of dozens of plans and the deliberations of numerous commissions over the years.

Varieties of phytoplankton that can be toxic cause 'red tides' that occur in many parts of the world. These result from nutrient or weather conditions that cause the algae to 'bloom' in billions, giving the sea a brownish or red appearance. Their toxins paralyse the nervous systems of fish, so they can't breathe – such die-offs can be massive. At times these toxins can enter shellfish like oysters, poisoning people who subsequently eat them. Those with asthma or other breathing difficulties should avoid beaches where red tides are occurring because the toxins from the algae can become air-borne, and are injurious when inhaled. Then there are green tides, like those afflicting the shallow bays of the Brittany coast of France, and which have put a $5billion tourist industry at risk. First appearing in the 1970s, these bad smelling green algae infestations are now an annual occurrence. As they rot they produce hydrogen sulfide, the 'rotten eggs' smelling gas that can kill as readily as cyanide. Their increase is attributed to fertilizer run-off from the land.

So much CO_2 is going into the oceans they are becoming more acid – according to the Oxford-based

International Programme on the State of the Oceans these infusions of carbonic acid have not been paralleled for at least 300 million years. More than a third of this carbon goes into the Southern Ocean, where acidification is already affecting tiny creatures called foraminifera, which need calcium in the water to grow their shells. There is also a problem with the tiny shrimps called krill, which swarm in the Southern Ocean and are a critical part of the food chain. There are an estimated 800 trillion krill in the sea – in spite of their small size their total mass exceeds that of all humans. Laboratory experiments indicate that if emissions continue to rise, acidity will increase so much krill might not be able to hatch in the Southern Ocean by 2050. Minke and blue whales, as well as a large range of sea birds, depend on krill for food.

As we rapidly exhaust the resources of minerals on land, robot machines are being developed to exploit metal ores in the sea, in water more than two miles deep. While seawater contains most minerals, they are in such small quantities extracting them directly is too difficult and expensive to be worthwhile. However, the minerals are concentrated in large quantities in two bizarre environments on the sea bottom. Where chains of active volcanoes extend beyond the coast eruptions from the sea floor create fantastic tower-like structures rich in minerals. Silver, gold, copper and zinc are among these, formed when very hot mineral-laden water emerges into deep seawater close to freezing point. These have accumulated into towers and chimney-like structures, sometimes as high as 200 feet, in regions where mineral-eating microbes might have been the first life on earth. Because no human could stay alive in these sulphur laden, toxic areas, bubbling with methane gas, robot machinery controlled from a ship on the surface will have to be used. Submarine volcanic events occur as the tectonic plates on which the continents rest move, creating not only earthquakes and tsunamis, but also volcanic emissions – magma, smoke, even geysers – miles below the surface. These have brought and will continue to bring minerals to the ocean floor – a perpetual metal factory. It is not yet possible to assess the quantities of minerals in these vents, since most are still to be discovered, other than that they must be very large. Two near Japan are estimated to contain metals worth $16million. In a field in the Bismarck Sea, off the coast of New Guinea, gold concentrations are as high as ten per cent, and zinc 20 per cent.

The second resource seems almost as improbable – a scattering of millions of billiard-ball sized 'nodules' with high concentrations of minerals across the deep ocean floors. The idea here is to collect them with tongs or giant vacuum-cleaners, and then suck them up through pipes to ships above, although there is also research into 'crawlers', robot tractors on plastic treads, which would collect the nodules, then crush them into a slurry that could be pumped to the surface. The nodules occur mainly on the abyssal plains more than two miles below the surface, and are most numerous in several parts of the Pacific and in the Indian Ocean. In places there are so many they touch each other, virtually covering the ocean floor. Consisting mostly of manganese and iron, they also contain other minerals in smaller quantities. One of these is cobalt, which is becoming rare in land sources, and which is essential to make the alloys in jet aircraft engines.

In 1994 the International Seabed Authority was formed to regulate and police deep ocean mining. Early attempts to mine either failed or proved uneconomic, probably because the technology was undeveloped. However, as the price of metals rose there was been renewed interest, and by 2014 the authority had issued 26 permits to mine in the central Pacific. Britain, China, Japan and South Korea all have projects planned. Not everyone approves. Even now, more is known about the outer planets than the sea deeps – 'It's tampering with ecosystems we barely understand that are really at the frontier of our knowledge base,' according to the vice-president of Conservation International, Greg Stone.

Oil is a major marine pollutant – more than 700 million gallons get into the oceans every year. But while tanker accidents make news, they are not major contributors. The largest single source is runoff from the land because of industrial activity and the improper disposal of used motor oil, which together account for perhaps 300 million gallons. Another 30 million gallons comes from tankers illegally washing out their tanks at sea. The hundreds of thousands of tons of plastic that find their way into the sea each year are mostly not biodegradable, simply accumulating in gyres and on beaches. On some estimates as many as a million seabirds and 100,000 whales, dolphins and seals are killed every year when they try to eat or are entangled in this massive debris.

Current fishing practices waste roughly a quarter of the world's annual fishing catch of about 90 million tons, causing the destruction of thousands of dolphins, turtles and other 'unwanted' sea life. Tuna often swim in close association with dolphin. When the less scrupulous commercial tuna fishermen see dolphin they put down nets to catch the tuna swimming beneath and behind them. The dolphin kill caused this way is enormous – at least 100,000 a year. Huge nets and 'long-lines', which can be as much as 80 miles long and carrying tens of thousands of hooks, are undiscriminating killers of marine and bird life.

Much of the sea is almost devoid of fish because the water lacks nutrients that feed the plankton at the bottom of the food chain – about two-thirds of marine life lives in the 2 per cent of so of the oceans that is nutrient-rich. So would it be possible simply to dump chemicals into barren regions of the sea and wait for plankton and the fish that feed on them to arrive? This seems an attractive idea on paper as a way to provide cheap protein to the developing world. However, some marine scientists don't like the idea, warning there could be unforeseen ecological changes, like a proliferation of toxic algae or reductions in oxygen that could result in the mass death of fish.

15 The Poles – More Important Than You Think

The sea is one of the vast elementals of the world we live in, but there are others, the north and south poles, influencing global climate profoundly as they drive a complex system of ocean currents, a huge conveyor belt moving warm water from the tropics towards the poles and shifting cold water back into the middle latitudes. If this system slackened or stopped altogether, the tropics would become intolerably hot and northern regions even colder. Much of the world now inhabited, like northern Europe, North America and Australia, would freeze or bake.

Polar vortices are windy pools of very cold air around the poles and polar jet streams are strong wind influences driven by the earth's rotation and temperature differences between the poles and the tropics. These normally operate as a 'fence' keeping the cold in, but they veer further north and south as the seasons change, especially in the Arctic, causing colder than usual temperatures in winter. In recent years unusual Arctic warming has weakened the polar winds more than usual, allowing pockets of frigid air and extremely cold weather to reach places that had seldom experienced them before – this is the most popular and perhaps the most plausible theory around for recent very cold northern winters. If it's true that global warming is the reason for more severe winters in the northern hemisphere, with the further implication that as the Arctic continues to warm, these cold winters may well become regular.

Climatologists suspect there may be other consequences. Hurricane Sandy is thought to have veered south towards New Jersey because of jet-stream aberrations. According to the UK Met Office in 2014 Britain 'had been affected very severely by an exceptional run of winter storms, culminating in serious coastal damage and widespread, persistent flooding... The clustering and persistence of the storms is highly unusual, for England and Wales this was one of, if not the most, exceptional periods of winter rains for at least 248 years.' The Met Office concluded there was 'a strong association with the stormy weather and perturbations to the jet stream.'

Until quite recently the five million square miles of Antarctica largely resisted human efforts to meddle with it – the desolation and extreme cold that characterize the continent seemed insulated, held in place by that formidable barrier of circular winds. The coldest, windiest, emptiest place on earth, there are no land mammals — polar bears are found only in the Arctic. Nobody had seen Antarctica till 1820, nobody went there until 1895, when the crew of a whaling ship landed. So apart from a few hardy adventurers and scientists, we are shut out by the most implacable weather in the world, where temperatures remain below freezing year round. This, it was widely theorized, meant the great southern region was unlikely to become a player in the huge happenings of global warming.

In 2014 this reassuring concept was abruptly shattered. A glance at a map shows a finger of land pointing up from the continent towards the southern tip of South America. This is the West Antarctic Peninsula, which is now very much involved, especially two of its great glaciers, Pine Island and Thwaites. These, debouching into the Amundsen Sea, were studied for eight years by teams of climatologists from NASA and the University of California, who reported in 2014 that these giant glaciers, which anchor much of the ice on the peninsula, were no longer held back by barriers of sea ice. This has melted in a warmer ocean, so the glaciers are now moving faster every year. The two studies concluded that collapse of the ice sheet in this region is now inevitable, and that this could contribute four feet to global sea levels, on top of three feet or more estimated from other causes. Beyond this, increasing instability in other areas of West Antarctic ice could raise the sea level potential to ten feet. There is no certainty about the timescale of any of this, but 100 to 200 years is theorized.

Other parts of Antarctica are raising some concern. After three years of observation by the European Space Agency's CryoSat satellite, which surveys 96 per cent of Antarctica the UK Centre for Polar Observation reported in 2014 a loss of 159 billion tons of ice from the continent in a year – a third more than eight years previously. While 87 per cent of this ice loss was coming from West Antarctica the rest of the continent was also contributing, with an ice loss of 26 billion tons. Just how much overall Antarctica ice cover will change in the future is uncertain – there are further questions arising about its stability. The Climate Change Research Centre at the University of New South Wales has identified much warmer seawater in East Antarctica, which 'could lead to a massive increase in the rate of ice sheet melt, with direct consequences for global sea level rise.'

Fifty nations have signed the Antarctic Treaty, which sets the region aside as a scientific reserve. Mining has been banned there since 1998, although China, which signed the treaty in 1983, hinted in 2014 that mineral exploitation there may happen eventually. China has five bases in Antarctica, runs an icebreaker, the *Xue Long* and devotes an annual budget of $55million to research in the region.

The Arctic presents much larger and more immediate dangers and opportunities. Because global warming is already reducing its ice cover rapidly, much that was once inaccessible is now open to humans, allowing mining for metals, drilling for oil and gas, and shipping movements. It is not known what the consequences of these interventions will be in what is already an unstable environment. This current degree of interest in the Arctic is something new – for as far back as we can trace human history it was shunned by humans as so cold and hostile it was best avoided, scarcely even thought of. It was conventional wisdom that human civilizations were substantially located in the middle latitudes – Aristotle felt human life could be advanced only in temperate regions 'between the torrid south and the frigid north.' Nothing had changed as late as the mid 19th century, when the hostility of the 'frozen north' was confirmed in the public imagination by the terrible fate of the expensive and highly organized Franklin expedition to seek a north-west passage. Nobody survived the loss of its two ships. Although there were several earlier claims by explorers to have reached the North Pole, it was not until 1926 that an international group headed by Norwegian Roald Amundsen crossed over it in the airship *Norge*. The flags of several nations were thrown out before the 300 foot airship was overwhelmed by bad weather and forced to land on the Alaskan coast, where it eventually disintegrated.

As the 20th century progressed, so did interest in the Arctic. Realization that traditional sources of minerals, especially oil and gas, were finite and the increasing accessibility of the Arctic because of global warming, prompted speculation that huge riches there were awaiting exploitation. The Soviet Union under Stalin began programmes of exploration and scientific research, maintaining floating ice stations from the 1950s on. Its Arctic territories were developed brutally, using the thousands of prisoners condemned to the penal settlements known as *gulags*. Two major mining projects, at Vorkuta and Norilsk, which were largely worked by prisoners, were above the Arctic Circle and the third and most important, the Kolyma goldfields, partly so. Kolyma produced 75 tons of gold a year at its height. Magadan, its urban centre, with 70,000 people in 1939, together with Vorkuta and Norilsk are still functional cities in today's Russia.

The Russians launched the first large ice-breaking ship, the *Josef Stalin*, in 1938, beginning a series culminating in the construction of the world's largest, the 75,000 horsepower *Arktica*, which in 1977 became the first surface ship to reach the North Pole. However, somewhat earlier, in 1958, the American nuclear submarine *Nautilus* had travelled under the ice at the pole. In 2007 two Russian submersibles placed a Russian flag inside a titanium container on the seafloor under the pole, more than three miles down. This was immediately seen as a Russian attempt to assert sovereignty. Claims came from other nations, including sovereignty over the North-west Passage. The Canadian Parliament, in a resolution passed unanimously in December 2009, went so far as to rename the route 'the Canadian North-west Passage, a part of Canadian internal waters.' This has also been contested.

Attempts to find a short northern route between Europe and Asia may go back to the 15th century, when a royal command dispatched Englishman John Cabot on a mission to America that might have included a search for a north-west passage. Cabot did not find one. Records from that time are fragmentary, sailing directions were scarce and often wrong, and conditions were especially cold because of the Little Ice Age. Many that followed fared no better, until in 1903 Roald Amundsen made the first successful passage in the tiny fishing sloop *Gjoa*, setting out in 1902. The journey took two years, with the boat trapped in winter ice for many months.

However, ice loss during the 20th century has made the northern polar routes much more accessible. The specially reinforced supertanker *Manhattan* made the passage with the assistance of two icebreakers as early as 1969, but the route was not considered cost effective. Quite a few yachts and other small craft followed, and in 2006 the cruise liner *Bremen* passed through. Perhaps the most spectacular passage has been that of the huge 'residential' liner *The World* in 2012. This ship made it in 28 days, carrying 481 passengers, including many owners of the 165 residences that make up this ship.

The five million or so square miles of the Arctic – – about the size of Russia — can be defined in two ways. The more arbitrary is to include everything north of the Arctic Circle, a line around the world at about 66 degrees north latitude, this being regarded as the southern limit of the polar night, the months of complete darkness during which the sun never rises above the horizon. Somewhat more practical is to regard the Arctic as those places where the average temperature never rises above 10 degrees C during the warmest summer month, July. Quite a lot of this land is below the Arctic Circle, within the group of nations whose territory encircles the Arctic Ocean. These eight nations, the US, Canada, Russia, Finland, Iceland and all three Scandinavian states, are members of the Arctic Council, formed in 1996 to protect and co-ordinate policy for the northern polar region. While the council has attracted praise for identifying areas of concern it has only limited influence, partly because it can do no more than recommend action, and partly because it is inadequately funded. Considering the huge changes now taking place, increasing military activity in the region by some council members, and rapidly increasing pressure to exploit the Arctic, this 'toothless tiger' status hardly seems satisfactory. The member nations can leave it as it is – a sort of *de facto* smokescreen over activities that might be less than scrupulous — or reshape the council as a genuine international force, able to intervene decisively if a mining venture, for instance, were damaging to the environment. There seemed some hope for this in 2015, when the United States took over the chairmanship. US Secretary of State John Kerry said he planned to refocus the council's efforts towards combatting climate change and safeguarding the Arctic Ocean.

The Arctic is already being industrialized and in no small way – also controversially, because of impacts on what has been pristine wilderness. Iceland's largest dam was completed in 2008 to provide cheap electricity to the multi-national company Alcoa to smelt aluminium. Thousands of foreign workers built the $1.3 billion Karahnjukar Dam and hydroelectric generator – employment opportunities for Icelanders were minimal. Almost all of the 690 megawatts generated goes to the Alcoa smelter, which produces almost a thousand tons of aluminium a day from bauxite shipped in from Australia. It was Iceland's third smelter. The project has been criticized on environmental grounds – increased erosion of riverbanks, silting of lakes, damage to fish and bird life, and contamination of the surrounding area with fluorine. Its apologists argue it is better for the world to produce the power-hungry metal with clean Icelandic electricity than that from coal-fired stations elsewhere. However, falling world aluminium prices put plans for a further major expansion on hold. In 2012 Alcoa said it would not go ahead with another smelter in northern Iceland, to have been powered by geothermal energy, citing not enough cheap power as the main reason. Iceland, a highly volcanic region, already gets a third of its electricity from steam coming from thermal boreholes. However there is a more ambitious project, known as Deep Vision, sponsored by the Iceland Energy Authority, Alcoa, Norway's StatoilHydro, and other foreign companies, which is drilling three miles down into the earth to tap magma at temperatures up to 600 degrees C. The idea is to create superheated steam, which could have a power output ten times that of sub-critical steam – to smelt even more aluminium.

There was some acid criticism in the United States Senate when America paid $7.2 million for 'a polar bear garden,' but there is little doubt now that the purchase of Alaska from Russia in 1867 has paid off. That pay-off began within a few years of the deal with the discovery of gold in many parts of Alaska, first at Sitka in 1872, followed by a major strike at Juneau. This gold rush brought thousands of prospectors from all over the world. One of the most extraordinary finds was at Nome on the north coast, where the gold was prolific in the sand on the beaches. More than 100 tons of gold have been gathered at Nome, which at the height of the rush had 20,000 inhabitants.

Gold is still being mined in Alaska, but it has taken second place to oil. This comes from one of the dreariest and least hospitable places on earth, Prudhoe Bay, located at 70 degrees north, where Alaska's flattish North Slope merges imperceptibly with the Arctic Ocean. It gets so cold at the town of Deadhorse – as low as -52 degrees C – that car engines must be run continuously to stop them seizing. This region has been producing as much as a quarter of America's oil since the field was discovered in 1968, but on some estimates as much as 30 billion barrels of oil remains to be discovered in Arctic Alaska, as well as six trillion cubic metres of natural gas. This would be enough to meet US needs for several decades. However, continued exploitation of oil resources, especially in the Beaufort Sea off the North Slope, is controversial for environmental reasons. Drilling uses dangerous heavy metals like cadmium and mercury, and there is always the risk of oil spillage into the sea.

This controversy was heightened in 1989 when the fuel carrier *Exxon Valdez* struck a reef in Prince William Sound, spilling over 11 million gallons of crude oil that contaminated 1000 miles of the coastline. A government study made 25 years later concluded that most of the natural systems damaged then had not recovered. The losses include the once prosperous herring fishery, which has closed, and the destruction of a large pod of orcas. Thousands of gallons of oil still pollute the beaches.

The Arctic weather has proved more than a match for oil majors hoping to find undersea reserves in the region – a major Norwegian investor, Jens Ullveit-Moe, calling Arctic energy exploration 'a license to lose money.' Halving of the oil price in 2014 made many projects there unprofitable. Shell announced a pause in its efforts to drill near Alaska's Kodiak Island after its Kulluk drilling rig ran aground in 2012 while under tow – according to the US Coastguard this was due to the company's 'inadequate assessment and management of risks in icy storm-tossed waters.' Although the rig had 150,000 gallons of diesel fuel and lubricants on board the fuel tanks remained intact in spite of a battering from 50 mile an hour winds and seas as high as 18 feet – Shell had just completed a $300 million refit to harden the rig to Arctic conditions. However the Russian oil and gas monolith Gazprom began production from its Arctic 30 platform late in 2013, and continued in spite of a demonstration at the site by Greenpeace activists, one of whom, Faiza Oulahsen remarked: 'This is a dark day for the Arctic. Gazprom is the first company on earth to pump oil from beneath icy Arctic waters …it is impossible to drill safely in one of the most fragile and beautiful regions of earth… we must stop this trickle of Arctic oil before it becomes a flood. There is no proven way of cleaning oil spilled in ice, and even a small accident would have devastating consequences on the Arctic's fragile and little-understood environment. ' Gazprom has signed a deal with Shell for oil and gas exploration in the Arctic, but Shell has said it won't proceed in the foreseeable future.

Alaska assumed great strategic importance during the decades of 'cold war' between the US and the Soviet Union, when nuclear war was considered possible. Prior to the development of long-range ballistic missiles these weapons would need to be carried by aircraft, strategic bombers, and the shortest way to do this was across the Arctic. In 1957 the US, with Canada, built the DEW (Distant Early Warning) Line of radar stations stretching from Alaska to Baffin Bay, and stationed fighter aircraft at Thule air force base in Greenland and in other northern locations in Canada. But it was only a matter of years before this concept was outdated by the development of inter-continental ballistic missiles and nuclear-capable submarines.

Two events, the end of the Cold War and the collapse of the Soviet Union greatly reduced the strategic importance of the Arctic, but at the same time revealed the extent of pollution the region had suffered. The Russian Northern Fleet was affected severely by years of financial stress. Serving officers went unpaid for months and there was little money to maintain nuclear submarines and the dangerous stores of spent fuel. In some cases nuclear material was dumped in the sea – there has been significant contamination of the sea off Murmansk. As recently as 2011 rusting hulls of submarines beached inside the Arctic Circle at Olenya Bay on Russia's Kola Peninsula were photographed. Of 170 nuclear capable submarines decommissioned there are official records of only 50 being dismantled. An unknown number are said to be 'mothballed' at various locations. In 2011 the Russian nuclear corporation Rosatom disclosed some details of the nuclear material dumped during the Soviet era – 17,000 containers of nuclear waste, 19 ships containing nuclear waste, 14 nuclear reactors, of which five still contain spent fuel, and 735 items of contaminated machinery. A complete large submarine, K—27, with fuel still in its two reactors, was scuttled in the Kara Sea in 1981 — there are fears these reactors may become critical again, and explode. This is not regarded as a complete list. A multi-billion dollar facility to store disused reactors and other waste began operating at Saida Bay on the Barents Sea in 2013, largely financed by international contributors. Fifty-eight reactor units were stored there that year – they must be held for at least 100 years until their radiation has subsided enough for them to be cut up for scrap.

The largest artificial explosion in human history took place in 1961 when the Soviet Union detonated its massive Tsar Bomba nuclear weapon over the Arctic islands of Novaya Zemlya. The statistics relating to this are mind-boggling. The explosion was the equivalent of 57 million tons of TNT, 1400 times as powerful as the Hiroshima and Nagasaki bombs combined. The bomb destroyed everything within a 22 mile radius, distributed deadly radiation up to 60 miles, and cracked windows in Norway and Finland as far as 500 miles away. The flash of light from it was visible for 600 miles, the 'mushroom' cloud rose 40 miles high, and the shock wave encircled the world three times. Although this bomb was modified to substantially reduce its fallout, there seems to be no record of how much there actually was, other than an estimate that 80 per cent fell worldwide. The most intense fallout was plainly in the Arctic Sea around Novaya Zemlya. There were 223 other nuclear explosions at this site, with a total yield of 265 million tons of TNT equivalent. Radiation pollution of the Arctic from these must have been considerable – however, in 1996 the International Atomic Energy Agency concluded that risks were high only in regions close to the dumping sites, and that doses to people and marine fauna were 'small and insignificant.'

And there are, of course, around 160,000 indigenous people living permanently in the Arctic. Once known as Eskimos – a word now seen as pejorative because it means 'eaters of raw meat'– they make up a group of associated cultures known as Inuit. That word simply means 'the people.' Their numbers were greatly reduced when they became exposed to European diseases, becoming a small, impoverished minority. In the 1970s enough Inuit were educated sufficiently to form political associations, which coalesced into the Inuit Circumpolar Council, now represented on the Arctic Council. The retreat of the Arctic sea ice has made climate change a major issue and has brought about an alliance with people on low-lying islands threatened by sea level rises – 'as our ice goes, so your water rises.' The Inuit peoples, once skilled hunters and fishermen, are now being increasingly absorbed into Western culture. We owe the word *kayak* and the small sea-craft it describes to the Inuit people, who used these canoes for hunting.

Global warming means the Arctic is now generating icebergs at an unprecedented rate. Greenland's massive glaciers are among the fastest moving in the world, spawning as many as 40,000 bergs a year into the sea at Baffin Bay. Some of these are carried south by the cold Labrador Current to the north Atlantic's busy shipping lanes – perhaps one in a hundred reach 48 degrees north. This greatly feared progression is called Iceberg Alley. Most of the bergs are small 'growlers', but a few weigh many thousands of tons, comparable with the monster that sank the *Titanic* in these foggy waters in 1912. However, even these growlers and the slightly larger 'bergy bits' have enough mass to sink a ship – each cubic metre of ice weighs a ton – and because they are mostly submerged they are much harder to see. The European Space Agency provides a satellite watch that can warn ships of icebergs in their vicinity, and the International Ice Patrol uses air patrolling and radar to check their movements. Nevertheless, there were almost 60 ship strikings in the 25 years to 2008. In 1959 the Danish passenger and cargo ship *Hans Hedtoft* hit an iceberg, killing the 95 people on board, and the 150 passengers in the cruise liner *Explorer* had to abandon ship for life rafts when she sank in a similar incident at dusk in 2008.

PART TWO: The Things We Do

16 Zetetics – and More

Zetetics. We all know that language is constantly changing, evolving, but few words are gaining meaning as fast as this one, and few have so many mysterious connotations. Several dictionaries fall back on a definition from the 1913 Websters – a dictionary put together a century ago –– 'a branch of algebra which relates to the direct search for unknown quantities.' To non-mathematicians this is a bit esoteric. For a time the Flat Earth Society (there really is one), cottoned on to the word for unfathomable purposes, and it is used by a variety of other occult and magical organisations. It can mean a study of the relationship between art and science, while a 'zetetic' can be a scientific sceptic who studies the paranormal with an open mind.

However, the Oxford Dictionary steers zetetics in a different direction and on to firmer ground: 'proceeding by enquiry,' and from here on definitions on the record give rather more depth – 'honest enquiry', 'a study of the science of research,' 'discernment based on logic, reason, and critical thought.' The world certainly could do with more of this – too little goes on around our planet that is informed by reason, logic and critical thought, as even a brief look around shows.

Our financial institutions seem to lurch from one crisis to another, most young women are compelled to work for economic reasons although respectable research has shown continuous contact between mother and infant during the first year of life is essential to proper brain development, we persist on spending huge sums on war and weapons, the majority of humans are impoverished, ill and barely educated, we seem incapable of dealing with an oncoming crisis as obvious as climate change.

During the 12,000 years since the end of the last ice age we have learned a thing or two, but how much of this has been wisdom, and how much just cleverness? It seems we need to think a lot more about our own behaviour, get ourselves together as a species, if we are to cope with the pervasive hazards this book has discussed. This need is so obvious, so important, it pushes me into the following attempts to sketch how some things might be better done. In doing so I will try to brace up to those austere demands of zetetics.

In two previous books, *The 2030 Spike* and *A Short History of the Future* I asserted that two axioms – self-evident truths – must be taken into account if any attempt at changing the human condition is to succeed. Over the ensuing years they have gained importance rather than otherwise, so I intend to put them forward again here.

The first is: *Useful change is likely to come only if it can provide as demonstrable, equal and general benefit as possible to the community in which it is planned.*

It could be put another way: if you do something, and someone loses from it, they won't like you, or it. Almost always they will fight back, a certain number will win. This is a very inefficient way of going on, and it gives no certainty that the best result will emerge. Unfortunately much of the recent debate about the future has been just this kind of confrontational struggle.

Our default position seems to be to assemble into two gangs, each putting an opposite, often extreme, view, and disinclined to listen to what the other side is saying. The current debate about climate change is typical – the deniers and the true believers put points of view so opposed they could almost be on different planets. The machinery behind this is typically 'cherry-picking', the selective use of only those facts that suit your argument. It may be an interesting game, but it is lethal to the truth and it wastes appalling amounts of time and effort.

Many nations still use adversarial trials to decide whether a defendant is guilty or not. These are a direct descendent of ancient trials by ordeal, and typically involve professional pleaders who put their side of the case, often their side of the case only. This makes it difficult for a jury to decide what the facts are, and also, sadly, gives a better chance of acquittal to those who can afford the best lawyers, who tend to be expensive. More enlightened states use 'inquisitorial' first hearings designed to establish the facts of the case.

Legislatures often follow the same confrontational pattern. Instead of the simple democratic idea that the people elect representatives to decide on public policy, those elected assemble themselves into two 'parties,' which again typically put adversarial views. Numbers become critical to the decision making process. The simplest electoral systems –'first past the post' – give power to those who get the highest vote, leaving a significant minority not represented in that legislature by someone they voted for. Again, policy differences are decided by a majority vote.

I once listened to a group of Indonesian graduates talking at a party in Djakarta about Western advisers urging 'representative democracy' on their country. They were not at all happy with what they called 'government by a half plus one.' For thousands of years village societies in Indonesia, as in many other countries in the 'under-developed' world, have decided matters of community concern by consensus – reaching a decision by discussion until a result is achieved that is acceptable to all concerned. Typically, this means everyone involved has probably given up some ground.

'Messy, and it takes time,' one of my Indonesian friends said to me, 'but it gets the better result every time.'

Israel's current plight illustrates the consequences of failing to observe Axiom One. There were without doubt many excellent reasons for the foundation of this new nation in 1948. The Jews, who had suffered such savage persecution from the Nazis, had no homeland. It seemed to the international community that there was every reason for them to have one, and since the Jews themselves believed that Palestine was their chosen country granted to them by God, and because some struggling Jewish co-operatives– *kibbutzen* – were already established there, it seemed an obvious choice.

However, people who were not Jews already lived there. There are plenty of Palestinians in exile who still have in their pockets the title deeds to land they say Israel usurped, and on which Israelis now live. Tens of thousands more are refugees. These people are not inclined to forget or accept this easily. Their disadvantage was a breach of Axiom One. The consequences? Decades of war and struggle, thousands of deaths, a perpetual state of tension throughout the Middle East, the dangerous acquisition by Israel of a nuclear arsenal, the steady transition of that country from a peaceful state to an aggressive one, and a permanent threat to world peace.

This is a heavy price to have paid, and it may well become heavier. And yet it could have been avoided. At the time other nations were willing to accept the Jews – a homeland in a better place than the strip of sand and stone they chose might have been found. Even now South American nations like Uruguay have huge areas of fertile, well-watered land awaiting development, and needing someone to provide capital. The largely empty Australian Kimberley region was suggested at one stage. The beautiful and productive Atherton Tableland in north Queensland has room for many more people – it is more spacious than many of the world's nations.

One doesn't have to look far to find other examples of offences against Axiom One - the British decision to favour the Sinhalese over the Tamils when

Ceylon became a new nation, Sri Lanka, the splitting of the Pashtun people of Afghanistan and Pakistan between these two nations, the serious disadvantage of the Kurds, a large people without a state – all these have caused thousands of unnecessary deaths and endless trouble. There is plenty of evidence that ignoring Axiom One doesn't only attract trouble, it pretty much guarantees it.

The idealist approach – that people should be good, honest, reliable, compassionate and thoughtful toward others – is undeniably worthy, but regrettably unrealistic so far as the majority of humans are concerned. Most people seem to be motivated by other things, their own self-interest, a desire for security, a fear of the radical and unknown. As a result all proposals for change should recognize this formidable aspect of human nature, which we can state as Axiom Two:

If proposed solutions don't take the lowest common denominators of human nature realistically into account, they will not work.

Self-interest is one of the mainsprings driving human effort – it cannot be argued away or ignored. Adam Smith, in his pioneering 1776 classic in economics *The Wealth of Nations* put it this way: 'It is not from the benevolence of the butcher, the brewer, or the baker that we expect our dinner, but from their regard for their own interest.'

It is one of the commonest statements about human beings that we 'ought' to be better, to be motivated by common decency, respect for others, the plain demands of the greater good. This is, of course, splendid, so we ought, but don't use this as a basis for public policy. That has to be pragmatic, hard-headed, and informed by an accurate estimate of what people are like, not what they think they are like.

For the purposes of this book I propose a third axiom: *There is a limit to how much economic, industrial and population growth the planet can support.*

There are plenty of people, many of them influential, who would deny this as a self-evident truth. When reminded that our natural assets are finite, some of them propose that having worn out this planet, we simply fly off to another one, and 'terraform' it to suit ourselves. Now zetetics solidly firewalls this kind of nonsense with the undoubted facts – no habitable planet is known, certainly none we could reach in anything less than several human lifetimes, our bodies and minds couldn't tolerate the stress and radiation levels involved, and even if we did arrive conditions would almost certainly be fatally hostile. According to scientists at the US National Space Biomedical Research Institute space is 'a sick environment, seriously affecting human muscle tone and the circulatory system, and exposing humans in space to huge dosages of cosmic rays, which cannot be stopped or avoided, and which cause genes to mutate.' It is more realistic to see Earth as a probably unique Paradise among a myriad of environments that are terrible for human life.

Given this, it is a logical conclusion that we should cherish the Earth, which we are not doing, and adjust our affairs so we can live sustainably on it, which we are not. Instead we tolerate planned obsolescence and its creatures, the appalling growth of 'packaging' and the throwaway society. Business seems to believe 'growth' is necessary so productivity can be maintained and they and their workers stay in business. They may also console themselves with the idea that if everything is left to 'market forces', the law of supply and demand, all will be well. Even though landfills are choked with plastic, and the sea marred with huge islands of floating debris, everything must be enclosed in a 'bubble pack'.

The consequences are familiar enough – short-lived devices, with service and spare part costs so high it is easier and cheaper to throw the thing away and buy another one, gadgets of illusory appeal, but not much use, too much stuff for the rich and not enough necessities for the poor. I shopped recently for new cutting heads for a well-known brand of electric shaver. I was quoted a price just over 50 percent higher than the cost of a complete new shaver of the same make, including cutting heads, which I eventually bought at a nearby supermarket.

This will be a familiar experience for most people, especially if they need spare parts for a car. The real issue is, should we tolerate this? And if not, what should we do about it? One possibility could be a state-run enterprise designed to finance competing manufacturers to make these bits and pieces if the original maker is guilty of rank exploitation. It would probably only need to do this very seldom – the mere threat could be enough to keep the offenders in line.

Surely there are remedies for these crazy and dangerous things that afflict our world – maybe just a matter of applied zetetics? Maybe. So why do people pay hugely more for, say, designer sunglasses, when something just as good can be bought much cheaper? Why do people run up their cards until they go bankrupt? Why is there such a demand for anything new, so something else perfectly serviceable has to be thrown away? Think kitchens. The answer is, of course, people, consumerism — shop until you drop. Now if resources were limitless, we could indulge all this. However, they are not, and we know, or should know, that we will eventually pay heavily for these extravagances. Is the answer to change people? Do we have the wrong approach to baby care, education, crime, business, money, and is that why things seem so prone to get screwed up? Maybe, maybe not, but it's worth looking at.

But how to start? I learned this the hard way when years ago I built a stone wall. My first approach was the lazy one – to take each stone as it came to hand and bung it in the wall. The project was a disaster, rickety and prone to falling over before it was half done. An elderly neighbour came to have a look. 'That'll fall down after the next decent rain,' he said. 'Surely you know you have to get the beginnings right, you need to be very sure the foundations are sound. The lowest stones need to be the biggest ones, they need to be the right shape, set well down into a good solid concrete footing.' So I started again, the wall got built, and now it looks as if it's always been there and always will be.

If you're constructing anything – especially a person – the beginning is the most important element.

17 The Next Generation

What goes for stonewalls also goes for people. Evidence is now mounting that what happens at the very beginning of life, the days and weeks immediately after birth, has profound effects on the rest of that life. Why are some people relatively calm and happy and others continually plagued with misfortune, at odds with everything? This is another area frequently seen simply as 'Life's like that'. But is it?   'Spare the rod, spoil the child.' This injunction to beat children has been around for a long time. The idea seems to come up first in English in William Langland's *Piers Plowman* in 1377, although Proverbs 13:14 in the King James Bible does assert: 'He that sparest his rod, hateth his son; but he that loveth him chasteneth him betimes.' This apparently divine authorisation of almost universal spankings conveyed definite religious overtones into physical chastisement in the 19th century, when it was commonly believed that a thrashing was needed to beat the devil from a child. That hiding was frequently accompanied by the pious hypocrisy: 'This is hurting me more than it hurts you.' It was mostly boys who got whacked. Girls were sent to bed and effectively starved by giving them only gruel to eat – a very thin soup-like porridge.

Opinions have changed with time. Although whether or not to spank is still an issue, serious research now indicates that copious love in babyhood and infancy, better still, extravagant amounts of love, have beneficial effects on a child's development, and that this lasts throughout life. American researcher Joanna Maselko of Duke University, who assessed interaction between 482 eight month old babies and their mothers found that 10 per cent of the women showed low levels of affection, 84 per cent 'normal' amounts, and six per cent high – caressing and extravagant. All the babies were then followed up for 30 years, the researchers assessing the reactions of these adults to stress, hostility, anger and anxiety.

Those few children whose mothers gave them the most extravagant level of affection came out best at handling all types of distress, especially anxiety –'High levels of maternal affection are likely to facilitate secure levels of attachment and bonding, which then translate into lower distress levels in both childhood and adulthood.' This resulted in lower levels of depression in young adults, with strong reserves of self-esteem and ability to cope with distressing circumstances.

Newborn children 'cared for,' just being fed and changed without getting any individual attention in Rumanian orphanages in the 1960s were found to develop a high degree of autism. They would repeatedly rock or bang their heads, developed problems with comprehension and had abnormally small heads. Follow-up studies showed that orphanage-raised children were twice as likely to develop mental illness than those in foster care. Only when research into this was published in 2007 did the Rumanian government change its policies.

These are among a number of studies that have established the importance of maternal bonding with newborn children, a link that researchers say is essential to the future development of the individual, especially for the successful establishment of intimate relationships as adults. In the words of another study group (J. Segal PhD and J. Jaffe PhD helpguide.com) 'You were born pre-programmed to bond with one very significant person… probably your mother. Like all infants, you were a bundle of emotions, intensely experiencing fear, anger, sadness and joy… The emotional attachment that grew between you and your caregiver was the first interactive relationship of your life, and it depended on non-verbal communication. This bonding determined how you would relate to other people throughout your life.'

Brain imaging technologies have shown these effects are not simply psychological – they influence the physical development of the infant brain in ways that are permanent. If this development is stunted, future intimate relationships may fail, with lower resilience to adverse events, an inability to maintain emotional balance.

Unacceptably high rates of youth depression and suicide are being reported from most parts of the industrialized world. For the past half-century it has become more common for women, including young mothers, to be in the workforce. While there is no absolute proof that these two things are linked, it may be prudent to consider this possibility. Try zetetics again:

experts warn that failed bonding in infancy has serious, fundamental consequences to the health and happiness of the future adult; there is conclusive evidence that these consequences are now more common than ever before; mothers are working more than before, and many do not have the opportunity to complete adequate bonding with their babies.

A primary duty for society should be to promote the successful nurturing and education of the next generation. Probably the most important part of this process is its foundations, the time of the earliest social impacts from birth onwards. Granted this, mothering emerges as a fulltime job, and a highly important one. Mothers need to fully understand the significance of bonding, and to be encouraged to give the time and effort to it – this applies specially to those who are obliged to work. Since nurturing is instinctive and comes naturally to most women, this should not be too difficult, provided they are not prevented by work considerations. Business does not always allow adequate paid maternal leave to working mothers without argument or equivocation – the economic element intrudes again here. 'Why should I pay you to have kids?' Quite apart from the obvious point that without them the human race would die out, such paid leave, provided either by the employer or the state, is a very good investment. Stable, happy people are productive, don't make much trouble and pay taxes. Maladjusted, unhappy ones cost the community money – a lot of money if their behaviour problems degenerate into criminality -- Chicago criminologist David

Anderson, who works in this area, has put the cost of crime in the United States at $1.7 trillion a year, while the cost of youth crime, people between 16 and 24, rose sharply to $1.2 billion in 2010 in Britain.

As recognition of the close association between inadequate early childcare and later criminality increases, most of the world's most successful societies provide generous maternal –and often paternal – leave as a matter of course. Sweden provides more than two years, shared between mother and father, Serbia one year, Canada almost a year, but at reduced pay. So, among other things, we ought not to smile indulgently at a young mother hugging her child and crooning 'Oh, you little bundle of joy!' – instead we might understand that such behaviour is a critical part of the framework within which the individual develops, the actual beginning of education. As many women would say: Every mother knows this.

Next, when considering education from infancy on, we may wonder whether, after getting through the complex and risky business of babyhood, these small individuals are suited to a 'one size fits all' pattern of teaching? And is that teaching as good as it might be – for instance does it make sense to teach literacy and numeracy to four to six year olds when the experience in several countries is that they learn these things much quicker and more easily at eight or nine?

It could be argued that learning to handle personal relationships, a broad picture of society and 'hands on' appreciation of the other life forms of the planet would be better priorities for early education. There have been some experiments in 'classroom democracy' to promote ideas like equal treatment, respect for others, that it is wrong to seek to dominate or bully another person. Respect for the planet and its huge variety of life might be promoted by hands-on activities like gardening and care for animals – this might also correct the idea that food comes from shops.

And surely it should be important at this early stage to identify, nurture and develop the natural abilities of each individual. Children often show an early interest in painting, music, dancing, tool using. It should be automatic for them to assume that they have a right to explore that potential, know what it is, and develop it. As things are, pressures to conform often stifle this early promise. Should a girl who shows outstanding ability as a violinist be forced to learn maths when all she will probably need in later life is enough arithmetic to count her money?

Traditional education systems, still current in many places, are based on ideas almost 2500 years old. The Greek philosopher Aristotle, in Book 8 of *The Politics* asserted that from the age of seven children should be educated to suit the demands of the state. This was a new idea at the time, when parents were teaching their own children what it seemed necessary for them to know. 'Education must be one and the same for all,' Aristotle wrote. 'The oversight of education must be a public concern, not the private affair it is now, each man separately bringing up his own children, teaching them just what he thinks they ought to learn.' Aristotle's model for schooling was adopted throughout the developing Western world. Schools assembled large numbers of children into 'classes,' where they were taught a variety of 'subjects' assembled into a 'curriculum.' Among these subjects until relatively recently were the 'dead' languages Latin and Greek, with which every educated man was expected to be familiar. Learning by rote was highly regarded – it was believed that memorising large slices of the 'classics', or poems, developed the brain.

It seems extraordinary that this system, which scarcely concerned itself with the needs and abilities of individual children, should have persisted as long as it did. At their extreme, these concepts saw the school as a 'sit-stillery' in which organized and largely arbitrary areas of learning – 'subjects'– were taught. Then, at the beginning of the 20th century, came an influential reformer, an American professor called John Dewey. Dewey believed societies could only develop by loosening up education, so children could form their own ideas and develop their natural creativity. Activity and experiment should replace rote learning and set 'subject material'. Seen this way, curricula became obstructive, even dangerous.

Austrian-born Ivan Illich was an ordained Roman Catholic priest until 1969, when he was relieved of his priestly duties because the Vatican disapproved of his ideas about education. Illich believed schools should be abolished altogether, and that teaching should become experience in real life situations. He vigorously attacked schools as dangerous, unbalanced institutions that did more harm than good. His alternative to them was a loose acquisition of useful experience from peers and elders, as well as practical experience with things – this seems to be pretty much what education was before Aristotle. Illich conceded that teachers would still be there, but as 'facilitators' in the learning process, which would not follow a curriculum, but would be designed to help individuals achieve goals of their choice.

Alvin Toffler remarks, in his *Future Shock,* 'Mass education was the ingenious machine constructed by industrialism to produce the kind of adults it needed… Our schools face backwards towards a dying system, rather than forwards to the emerging new society, which will require people who have the future in their bones.' This quote identifies the two conflicting objectives of education – to mould children to fit society and its purposes, or to allow free development of the abilities and ideas of young human beings. Social and religious prejudices are very much in play here.

Social systems in which the working class is expected to 'know its place', in which the 'deserving poor' are seen as inevitable, are regaining ground in many places. Since most women now work, schools are significantly places for children to 'be' while both parents are at work. And while many church schools, like others, are becoming more liberal, there are plenty that tend to be conservative, teaching what suits the dogma of that particular religion. This seems especially the case in the *madrassas,* the Islamic religious schools, in many of which rote learning of large sections of the Koran is mandatory, but not much else is taught.

What are Met schools? American teacher Dennis Littky, frustrated with what he saw as failures in education, retired to a cabin in the woods to think about it. Then in 1996 he tried something new. No teacher discipline, no exams or formal lessons – this was the new pattern established at the now famous Met school in Providence, Rhode Island, arguably one of the world's most successful places of education. In spite of the fact that about half the students were from poorer homes, and had limited previous learning skills, every one of the first two graduating classes were accepted into university. Three-quarters were the first in their family ever to go on to higher education. Less than half were white, 38 per cent Latino, 18 per cent African Americans.

In Littky's words 'The main thing is not to be boring.' Major objectives are to teach students how to learn and think, and to develop their individual abilities towards some purpose – a purpose they have probably decided on for themselves. The structure is based on groups of about 14, who stay together for four years with a teacher adviser, pursuing largely individual goals. This concentration on each individual is a strong theme. Some work best alone, some need more time than the average to get things done.

How do children learn? Research into successful home schooling shows they do this best by asking questions about whatever comes into their minds, and that these do not follow in any organised 'curriculum' way – 'Do cats think?' 'Why is the moon yellow?' 'Why is sugar bad for you?' The evidence is strong that children have a natural ability to learn this way, and then fit all these things together into a coherent picture of the world.

This raises all sorts of questions. To what extent are traditional schools still a product of the 'clockwork universe' ideas of the 19th and early 20th centuries, which dictate a mechanistic view of education as with everything else? To what extent does this suppress the natural abilities of children? To what extent does boredom act as an abrasive in the process of learning?

Curricula could become much looser, giving greater freedom for individual development, and in many schools this is already happening. Admittedly, this does make formal examination more difficult, but that should be secondary to providing children with the best possible education – if employers want qualifying exams, they should organize these themselves. Ideally, school time might offer students the freedom to select, from books and audio-visual material, the things they feel they need to know, and which contribute to their aims in life. There are already electronic inter-active sessions with teachers of proven ability and the best communication skills, even if these do verge on the charismatic – these could profitably become universal.

Once the viewpoint changes to giving children what they are likely to need in life information on how to buy a used car, get a good discount, how to get on with boys (or girls) float into the picture. Open-ended learning schedules, rather than fixed terms ending in exams, would produce students approaching excellence in their subjects in their own time. The present system, setting a pass mark that concedes it does not require full knowledge of the subject area, is a tacit admission that it generally produces half-educated people. Its extension into universities is most regrettable.

There are, of course, schools that recognize many of the principles outlined in this chapter. They tend to be labelled as 'progressive.' Such are the 995 Steiner-Waldorf schools operating in 60 countries around the world. They say they 'aim to help every child pursue her or his unique destiny, stressing the artistic and the role of the imagination strongly – seeking to provide young people with the basis on which to develop into free, moral and integrated individuals.' Steiner schools generally discourage kindergarten and lower grade children from being exposed to media influences like TV, computers and recorded music, which are seen as harmful to their development.

Whatever. One way and another, we have opened up the door on our first crucial area, how we might use the upbringing and education of our children more effectively to develop socially adequate, innovative, happy people who recognize their own abilities and ambitions and are confident in them. The challenges of the world not too far ahead will demand no less. Granted an oncoming generation who might achieve this, what sort of world framework will they need in which to operate efficiently? It will certainly need to be something better than the international jungle we have now, devoid of any binding rule of law – governed rather by a complex web of fears, prejudices and national interests – – in a word, anarchic.

18 One World or None?

Smallpox. Very few people know much about this horrendous disease any more, or even think about it, because, except for a few 'experimental' germ banks, it no longer exists. This was not always so – as recently as the late nineteen-sixties it was killing two million people a year. That was before 1967, when perhaps the world's most consequential and successful global enterprise began – the total elimination of smallpox with a programme of mass immunization. Smallpox was a miserable disease, which had plagued the world as far back as human history can be traced. Starting with flu-like symptoms, muscle pains, headaches, it killed most of the children and half of the adults who caught it.

The timing of this successful global achievement is significant, since it spans the era of the Cold War, during which the Soviet Union and the United States were on severely hostile terms, at times on the brink of actual war. In spite of this they were able to co-operate effectively *on a worldwide basis* in the smallpox project. It has been argued, and convincingly, that this happened because of the involvement of a third mediating and controlling body, the World Health Organization, which is an instrumentality of the United Nations, transcending nationality and able to operate on a global scale.

And, remember, this was perhaps the greatest single episode of disease control ever achieved, extending over more than a decade. If such a remarkable effort could succeed once, why can't others like it?

It is tempting to think about a global community with a common language, in which people who are citizens of the world move freely around, allied more by common interests than nationality, taking peace and prosperity for granted. In many ways the stage is set for that, the technology is rapidly becoming available. There are, after all, no essential reasons against the extension of the rule of law from the national to the global scene. A society without dissension, without crime, is hardly possible, but a reasonable rule of law should not be out of reach. Most countries have it – why not for a world that needs it so much? But we have the United Nations, you might say. As the smallpox epic shows, the UN and its instrumentalities can and do achieve a great deal of good.

The real point is what they can't do, and that is substantially due to a lack of teeth – a refusal by the nation states to give enough power to the world authority to make it fully effective. The reality is that the United Nations is often restricted by the deliberate withholding of funds, the veto powers of the Security Council, and a naïve and impracticable system of voting in the General Assembly, where every country gets one vote regardless of its size.

The governments of the major powers ignore the International Court of Justice – this means, among other things, that crimes against humanity, as in Sudan, generally cannot be prosecuted effectively. There is little agreement about what human rights are, or how they should be enforced. When 120 nations voted in Rome in 1998 for an International Criminal Court to punish those violating 'fundamental freedoms' the world's four largest nations, China, India, the United States and Indonesia, opposed it, and continued to do so when it was ratified by 66 countries and established in 2002.

The resulting world anarchy is a major hazard – whether or not we can remedy it must affect our future profoundly. As more nations get nuclear weapons and overall prosperity dwindles, the precarious balance of power in the world is fading. Less and less is done for the starving and the afflicted, there is no co-ordinated care of the oceans and the air, and the 'fortress' mentality of the West, Russia, Japan, and perhaps also China, will strengthen. The more unpleasant manifestations of climate change may well compel co-operation in time, but how much better it would be if we could work together on a planetary basis before this disagreeable medicine is fed to us.

But whatever its merits it would be naïve to expect a transition to world government quickly – most of those thinking seriously about it envisage a slow, progressive evolution, a time during which nations surrender some of their powers by careful multilateral agreement. But do we really want to do this? Almost without exception we are conditioned from childhood to believe in a concept based on occupation of a certain amount of territory – the nation. It must be the object of our loyalty, our respect, our devotion – we may even be called on to die for it.

This patriotism goes back to our primitive origins – many other forms of animal life mark out and defend territory. In most people it is so deeply ingrained they take it for granted, and can become actively hostile to anyone who doesn't.

During past millennia there has been a good deal of point to this, because humankind was indeed broken up into hundreds of small territorial units. It was normal for people never to move out of their own country in their lifetime, often enough never leave their home village, and to be almost totally ignorant about their neighbours. Virtually their only contact, apart from very limited trade, was when they were fighting each other, generally at the behest of their local lord. This happened so frequently it was a major factor in restricting the world's population. War has killed a lot of soldiers, but as many if not more civilians. Throughout much of history armies have 'lived off the land.' This forced seizure of crops and animals, and as often as not the burning of villages to deny them to the enemy, destroyed rural economies as the former farmers and their families died of starvation. All this happened in conditions where most people had almost no global consciousness – even kings and emperors believed in such mythic characters as Prester John, an eastern king with a magic mirror in which he could see anything going on anywhere in his vast dominions, and possessing other supernatural powers.

But the world is very different now. Millions of people go to other countries as tourists or to work, and huge and increasing numbers of refugees are on the move. The news media keep everyone informed from minute to minute about global events. The Internet spans the world, email gives instant communication everywhere and we can look at and talk to friends and relatives far away with Skype. Billions use the social media. This profound linkage, which is something totally new to the world, gives reason to ask whether nationhood is outmoded.

Let's look at the benefits first. Much of the charm and interest of the world comes from its huge range of cultures – we'd lose a great deal if there were the same kind of people doing the same sort of things in places not very different from anywhere else. People take pride in their language, their literature and art. And, as we have noted, the nation states are compact enough, and their people sufficiently similar in their outlook, to allow law and order to be maintained to a reasonable extent. In some places, not all that many, systems of representative government are maintained, and just a few approach actual democracy.

Then there is the other side – the disadvantages of the nation state. While states are sometimes invaded and taken over by others, the general idea is that there is something sacrosanct about nations. They must be respected and their leaders must not be usurped, no matter how badly they rule their people, no matter how inefficient the society is.

Is there a need for an effective international authority, with police powers and the muscle to maintain a coherent and just body of world law? Ask the women of Darfur, the conscripted slaves of Burma – and, for that matter, the starved and stunted children of North Korea, denied the very means of life to support the privileged, heavily regimented minority behind the military autocracy. And these are not the only 'rogue states,' there is a long list of others, and more crop up every year. The bottom-line is they get away with it because it is assumed their national integrity must not be usurped.

A large number of people are thinking about whether to establish a world government and if so, how? There are numerous websites claiming that conspiracies already exist for a sinister and clandestine world dictatorship. All kinds of candidates are proposed – Jews, Chinese, Moslems, black people, multinationals, churches, to name just a few. High on the list is the Bilderberg group, a secretive collection of politicians, businessmen and princes formed 50 years ago to consider world events. This group, without doubt very influential, is closed to the media and unwilling to disclose the results of its deliberations. Blogs and other web sites reveal a good deal of anxiety that a world state could be authoritarian, and not necessarily benevolent. Since more than 100 states torture people regularly, this fear is far from irrational. At least in the case of rogue nations it is possible to escape elsewhere – there could be no escape from an oppressive world government.

The facts set out so far then, from a zetetic point of

view, suggest a gradual and cautious approach, and a definite separation of powers. Whatever evolves must preserve regional cultures and ways of life, must use wide consultation of people on a world basis, have enforcement machinery that transcends national interests, eliminates war, torture and slavery and adequately protects common property such as the sea, the atmosphere, and the planet's flora and fauna. A code of conduct for multinationals, a reduction in spending on weapons, and a transformation of war industries to things we really need, like alternative energy infrastructure, would be desirable. These are worth considering point by point.

The first major consideration, that powers should be divided, means the world government would deal only with matters of international scope like world peace, care of the commons, human rights, climate change. Although it would be a mammoth task, an international electronic referendum to establish just what powers a global authority would have, how they would be implemented, and by whom, is not impossible. The issue would certainly be important enough to justify this unprecedented kind of consultation. Residual regional and local matters would continue to be handled by regions and countries, which would maintain their individual cultures, and make and administer local law with complete autonomy, except that they must abide by international law.

Since the trend is already towards global Internet usage, almost universal participation is not impossible and could well be made possible quite soon by natural growth of the Internet. Of three billion users in 2015 – 40.4 per cent of world population — by far the greatest number, almost half, were in Asia. Europe came next, with almost half a billion, North America third at slightly over a quarter of a billion.

Can war be eliminated? Many people have asked this question, and almost as many more have said it can't – they consider a basic instinct in humans to fight is too deeply ingrained. Yet most countries have managed to control open conflict inside their own boundaries, and people generally get through their lives peaceably enough.

It is naïve to suppose that peace could break out all over the world. The people and materiel of the armed services could be converted into a world police force, able to control the regional conflicts, piracy and international crime that are bound to continue. Their expertise and equipment might also be used profitably to help those increasing numbers who will be injured and dispossessed by the climate catastrophes of the future. Military discipline and equipment could also serve the next requirement – care of the commons. Practical means, including the use of force, will be needed to control such things as illegal over-fishing, dumping of bunker fuel and plastic waste into the sea, monitoring air quality and the consequences of sea level rise. Many people now employed in armed forces may well prefer to serve in these capacities.

None of this would happen quickly because it

could not – there is too much mutual fear between the nations for that. It would probably take decades of gradual transfer of powers before nations would agree to a total turnover of the means of force to an international authority.

When the oil-drilling rig Deepwater Horizon exploded and sank in the Gulf of Mexico in 2010, 11 of its crew died and millions of gallons of crude oil poured into the sea, caused massive pollution to the United States' eastern shorelines. The rig, technically regarded as a ship, was registered under a 'flag of convenience' in the Marshall Islands. According to Minnesota Congressman J. Oberstar: 'Coastguard inspection of a US-flagged mobile offshore drilling unit takes two to three weeks, but safety examination by a foreign flagged unit takes four to eight hours.' He seriously questioned whether the Marshall Islands registry, which is run by a private firm, could be responsible for ensuring compliance with quality standards for the construction, equipment and operation of the rig. That disaster has cost hundreds of billions of dollars.

Forty per cent of the world's shipping is registered in three tiny countries, Liberia, Panama and the Marshall Islands — in 2014 more than half of the vessels at sea had this kind of 'phony' registration. In those ships seamen are often paid badly or not at all, safety standards are poor, and the actual ownership of the vessels can be deliberately obscured to avoid liability. They are frequently involved in international crime, including the trade in drugs and in young women, as well as illegal and destructive fishing.

And the world needs other specialised agencies, dealing with issues in science like climate change and genetic engineering. There is a clear need for a balancing economic authority, with political and policing powers spanning the entire planet, to regain some of the billions of dollars in tax evaded by multinational corporations – why not a simple turnover tax, without exceptions, imposed at the same rate in all countries? Other global agencies could set standards for human rights, transparent government and reasonably equal opportunities for schooling and earning money.

In ancient Athens, generally regarded as the birthplace of democracy, citizens voted by dropping a black or a white pebble into jars. That was all right for a small city-state, but as nations grew bigger direct democracy was seen as impracticable and representative government was introduced in which we elect delegates to theoretically do what we want or need. That worked in communities where most people were illiterate and discussion of politics was largely confined to the few 'upper class' men who were entitled to vote. A consequence of this was the two party system, which offered an easy choice to an electorate, which neither understood politics nor wanted to. But in today's world being able to vote every three or four years is not enough for better-educated and vocal populations. This denial of participation is a big factor in the present growing disillusionment with government in major 'western' countries.

During the 20th century the 'western' system went through a heady missionary phase. 'Democracies' were set up all over the world, especially in the ex-colonies. They have not fared very well, many turning into autocracies, and not very nice ones, at that. Others are 'guided democracies,' which have elections and houses of parliament, but which are effectively governed by a single junta. Their architect was the late Singapore leader Lee Kuan Yew. That is the way Singapore is run, and its undoubted prosperity and order, as well as Mr Lee himself, are highly regarded in China, where there are those who see Singapore as a model for China's future.

In his lifetime Lee directly challenged Western democracy, remarking that east Asian communities in particular 'place more emphasis on order, stability, hierarchy, family and self-discipline than Westerners do. The individual has to realize that there are broader interests to which he or she must be subordinate.' He compared the economic failures in Europe with the orderly prosperity of Singapore.

One morning when I went to the library in the small American city of Eugene in Oregon State a group of perhaps half a dozen young people greeted me. They wanted me to sign a petition for an electoral initiative, which, not being a US citizen, I was not qualified to do. But they were ready to talk about the issue they were proposing, and the political scene in the state. I found them enthusiastic and well informed — as they said, because they could be part of the political process.

The machinery for the initiative and referendum, often styled 'people power,' exists in 24 American states and a scattering of smaller nations, including Switzerland and Scandinavia. It means that if a group of concerned citizens can get enough signatures on a petition — usually about five per cent of the electorate — the matter proposed must be voted on in much the same way as a general election. If it gets a majority this 'proposition' becomes law, quite independently from the legislature. This represents a significant shift in power from politicians to citizens.

This placement of power into the hands of American citizens was introduced early in the 20th century because of a perceived domination of state governments by corrupt influences. Often initiatives tackle big issues politicians have put in the 'too hard basket.' Universal military training in Switzerland was introduced by referendum and in 1997 another in Oregon established the option of euthanasia for those with painful and incurable illnesses — the first American state in which this became law.

George Washington didn't like the two party system, saying the constant alternation of two parties in the federal government would be 'a frightful despotism.' He refused allegiance to any political party during his eight years as America's first president. Two other early leaders in American politics, Alexander Hamilton and James Madison warned against party politics – they were afraid of 'a spirit of faction.'

When I first became a senator in the Australian

Parliament I had the idea, perhaps much like George Washington, that democracy meant representatives coming from their constituencies ready to freely discuss national issues, and then make laws for the common good. I was quickly disillusioned. Because of the way the numbers fell out my small party held the balance of power. This obliged all of us to work hard to understand every bit of legislation coming up. This was not the case with Government and Opposition members. Most of them contributed little to the parliamentary process, as they would sometimes wryly say, they were 'bums on seats', obliged to vote the ways their party Whips dictated. Quite often, as they filed in when the division bells rang, one would say: 'What are we voting on now?'

The two party system indeed means decisions are made by a few people while the majority do as they are told. This suits the media, the lobbyists and pressure groups, who only have to deal with a few people — a much easier process than talking to every individual member. However, it also represents a significant narrowing of opinion and is likely to give too much weight to the beliefs and prejudices of the few. And because the few are very busy, they can drift out of touch with public opinion.

19 A Place to Live

Why not make a small but strong and efficient house with a 3D printer — or maybe ten houses, a hundred, a thousand? In 2014 a Shanghai company built ten small houses this way in a day from quick-drying cement and demolition rubble, using four huge printers, each 33 feet wide and 22 feet high. Each 210 square foot building cost less than $5000. The same firm, WinSun, subsequently prefabricated a five-floor apartment building using the same methods. Should you regard this as just a quaint oddity, consider the following:

When a force 7 earthquake – a not especially severe one – struck Haiti in 2012 *250 thousand* people were killed and 190 thousand houses destroyed in the capital, Port-au-Prince, which has a total population of around two million. Why? Because '86 per cent of the population lived in mostly tightly packed, poorly built concrete buildings,' according to the Disaster Emergency Committee that investigated this catastrophe. Another earthquake about the same size two years earlier did extensive damage to buildings in New Zealand's second city, Christchurch, but only 185 people were killed, because better-built houses withstood the shock. Significantly, more than half of those deaths were in one structurally deficient high-rise building.

For billions of people home can be the Steamroller hazard that kills. More than half the world's population lives in dangerous houses, rickety, put together with frail materials and tragically vulnerable to fire, flood and earthquake. Concrete that is not properly reinforced kills when walls fall on people, bamboo walls and thatched roofs burn only too readily. This, then, is a large hazard that needs attention and if mass production of houses with 3D printers turns out to be one solution, why not?

Only three per cent of humans lived in cities in 1800, in 2014 54 per cent did. There are those who think this is a good thing, but is it? *The Economist's* 2014 Liveability Ranking reported that the quality of life in more than half the cities surveyed had deteriorated over the previous five years. Cities attract people because they are cosmopolitan, a lot goes on there, live entertainment and sport, art, crime, eccentricity. They even provide individuals with some welcome privacy. Lost in the crowds, you can be yourself. In a village everyone knows your business.

These are some of the attractions the great Western cities provided as they grew. As someone once said: 'If you grow tired of London, you are tired of life.' Once the world's largest accumulations of humanity, they now rank well down the list. The biggest, London and New York, with around eight million each, share eleventh place, outnumbered by the giant mega-cities of Asia and Latin America. Here life is not so idyllic. One in three city dwellers lives in a slum, and that number is fast increasing, according to former UN secretary-general Kofi Annan. By 2020 more than two billion people — at least a quarter of mankind —- seem doomed to live in informal squatter settlements and improvised slums that are neither recognized nor serviced by city authorities. People die young in such places – many of the millions of children we lose before they reach five come from them. Climate change, population pressures and strained food resources can only worsen this situation into tragic dimensions.

In most of the world's biggest cities, as many as half the people live in these untidy skirts of unspeakable slums. In Kolkata, in India, a third of the population literally live, eat and sleep on the streets, often forced to rent small areas of pavement from criminal gangs. Other cities with populations approaching or over 20 million with the same kind of problems are Sao Paulo, Delhi, Manila, Dhaka, Djakarta and Karachi — significantly all of these are in regions of rapid population growth. This mass disaster has happened because there are now ten times as many people in developing world cities than there were in 1940 – and these places are continuing their uncontrolled sprawl into the surrounding countryside. Many are expanding on to coastal land that will flood as sea level rises. The governments who try to administer them generally don't have the resources to keep the existing infrastructure in good order, much less build anything new. This ensures that most new developments are slums, with a high incidence of disease, especially tuberculosis, and the appalling crime rates of places like Mexico City and Manila.

Chongqing, in south-west China, which you may never have heard of, claims the biggest city size at 32 million, but this is because it includes neighbouring regions in its administrative area. It is certainly big and growing fast, but at about 9 million it ranks in the second row of megacities. The real monsters are Tokyo, 38 million, Delhi and Shanghai, 25 million, Beijing, Mexico City, Mumbai and Sao Paulo, all over 21 million and Osaka, 20 million.

Dense concentration of people and inadequate building standards have turned many of these cities into lethal traps when they are hit by even relatively minor flooding and smaller earthquakes. Unsafe apartment blocks were major contributors to around 40 thousand recent earthquake deaths in India and Turkey. Inadequate foundations and steel reinforcing, and too little cement in the concrete were found in more that half the damaged buildings in Turkey. Unfortunately for the world, Christchurch's relatively small death toll after its earthquake was an exception. Death trap buildings are much more typical, due to economic stress and inefficient governments, mostly authoritarian and often military.

As we shall determine, when we finally wrap things up towards the end of this book, one of the most dangerous things you may do as century 21 progresses is live in a city. They are highly vulnerable to natural disasters, and in time of war. Already lots of people are killed and injured on their traffic-choked streets — the bigger cities get, the greater the risks. Air pollution, so severe it represents a dangerous health hazard, has become characteristic of most major cities, especially the largest, and many have been forced to regulate the causes of pollution simply to remain habitable.

India's Delhi had the doubtful privilege of being the world's most polluted city in 2015, according to a World Health Organization survey. Its level of dangerous PM2.5 particulates was assessed at twelve times the recommended maximum. India's death rate from respiratory disease, 159 per thousand in 2012, was around twice that in China. Beijing, China's capital, is also severely affected – in 2015 its mayor, Wang Anshun, declared it 'unliveable' because of noxious smog. Although the government took extraordinary measures to clean it up for the Olympic games in 2008, shutting down industries and ordering cars off the streets, conditions have worsened since, with severe and foul-smelling mist reaching an almost intolerable level. A businessman began selling canned air from Tibet, and suppliers of air filters and masks could not meet the demand. Many other Asian cities are similarly afflicted, and at a heavy cost, with respiratory disease a leading cause of death.

Traffic deaths are especially high in and around the megacities, because of the generally chaotic state of the roads. In India in 2014 almost half a million road accidents caused 130,000 deaths – the world's worst traffic accident rate.

This is not a very good checklist. On the whole the Blue Planet must rank pretty low in the galaxy for urban development. So can we do better? Must we do better? Developing mass-produced sturdy but low cost pre-fabricated houses to replace the fire-prone and shoddily built shacks so typical of the villages and city slums would plainly be useful, but many architects and designers now concentrate on expensive structures. Computer assisted design and automated manufacture are permitting structures like Gehry's Guggenheim Museum in Bilbao, Spain, Libeskind's additions to the Victoria and Albert Museum in London, the National Museum in Canberra, but these costly buildings have little relevance to the world's poor.

Almost all big cities are now plagued by an essentially artificial factor, the high and increasing value of urban land. This means dwellings become smaller and closer together, and more isolated from nature. In Tokyo, where the majority of people live in tiny apartments, some businessmen sleep weeknights in capsule hotel 'rooms' not much larger than a coffin, braving the packed trains to go home only at weekends. This approaches the science fiction nightmare of humans conditioned to live in isolated cocoons, surrounded not by real things but by 'virtual' reality.

Such absurdities, as well as condemning young couples to work half a life time to own a small box in the sky, argue powerfully that even in the wealthiest Western societies our cities are becoming impracticable, near the end of their useful life. Looked at in the cold light of zetetics, cities cannot solve the basic, seemingly intractable problems outlined earlier in this chapter, so is it more reasonable to seek those solutions in totally new and different kinds of habitat?

If so, a high priority could be to get people out of

large cities into carefully planned regional complexes that offer living conditions less demanding than the cities, which are better suited to families, and which offer reasonable work opportunities. This is not a new idea. There are thousands of 'eco-villages' around the world, many successful, others failing from want of support. Their basic ideals are laudable enough — life in harmony with the land, conserving its eco-systems, organic, sustainable agriculture, co-operation and consensus politics. There are many reasons for their failure, but usually it's because they are too small to offer an acceptable range of employment and services and a sufficiently varied society. This is also why thousands of small towns and villages in most Western societies, and increasingly in developing ones, are languishing.

Business likes people to be in cities because business is there, and it is convenient to have the workforce, like other commodities, neatly packaged in boxes nearby. The urban jungles of today are indeed a product of industrialization. When mass production of goods developed, it was economically desirable to have workers conveniently near, so whole new suburbs, rows of terraced houses in many places, were built to accommodate them. High-rise apartment blocks are their modern equivalent.

It is worth asking whether this way of life is natural and sustainable, and does it have to be permanent? After all, it evolved quite recently in terms of human history. Before the industrial era men generally lived in the country or in villages, working the land on which they lived, along with their wives and families. This life was frugal, partly because half of the food produced went to feeding working animals, but it did have its merits. However, the traditional village is not really an option now, because community size is important. The challenge is to create urban areas that are optimal in size, providing lifestyles at least as attractive as those in large cities.

For most people, the work element is critical. Their job is in the city. If they don't want to commute long distances, it is easiest to live near their work. Then there is fun. The big entertainment and sporting venues, the places where people want to be seen, are usually in cities. Yet even these considerations are receding as the electronic media give you a better look at these events at home without the parking problems. Millions already work from home – telecommuting – without being close to, or even visiting, the 'office.' Outsourcing is creating millions of small contract businesses that are replacing the huge offices and factories of the past. For most people, electronic and social media are fast becoming the main diversions, and these can be enjoyed anywhere. Gardening is becoming popular – even vertical gardens are evolving for space-limited apartment balconies. These trends suggest smaller, less centralized living areas specialized to suit modern needs and preferences, with access to at least some surrounding land. Artists, writers, anthropologists, butterfly collectors, golfers may like to live close to one another. This would be much more feasible in purpose designed and built habitat than in the amorphous mass of a modern city – even the best of them.

For this reason and others the physical infrastructure of the megapolis is a second problem area. Many of these problems arise simply because almost all cities are old, can no longer cope with modern conditions and exploding populations, and don't suit modern lifestyles. Narrow streets block traffic, but the surrounding property has become so valuable it is uneconomic to widen them. Even if widening does become possible, traffic soon seems as bad as ever. Better public transport could answer this – perhaps our new habitat should have moving footways. ('Seats on them, please!' a lady friend comments) There are problems with other infrastructure. Sewerage, water supply, electricity, and gas – all these services become increasingly difficult to repair or extend to cope with rising populations, and they become relentlessly more expensive.

Perhaps the answer is a complex of say a dozen 'towns' with planned open space between them, good internal communications, and some specialization of activities. Health services might be in one town, shopping in a second one, the university in another. Recent development of Singapore, where nearly 5 million people are crammed into less than 400 square miles, has included some of these ideas.

These considerations lead us towards housing that does not have a high land cost, that provides what is needed for a natural and creative childhood, that encourages peer group activity among children, that can accommodate modern 'coupledom', that can help absorb the shock of relationship breakups, that can provide effective social reinforcement, with habitat of a high order and local employment, and which is energy efficient — a tall order, but not impossible. The modern imitation of the traditional village – satellite towns – designed to provide cheaper land, with more open space, has not been notably successful. Most have been mere 'housing estates', providing pretty minimal housing at that, with little thought for social infrastructure.

The older village concept seems more promising, but it must somehow be grafted into a complex that is big enough to provide services at an acceptable level and enough social variety to be interesting. This might involve a very different kind of 'household' from the one we are used to, designed for one man, one woman and their children. The new form of habitat could be a wide circle of large 'houses', each big enough to accommodate say, 20 adults and their children, each group within it occupying its own self-contained module, in comfort and privacy, but sharing facilities like a crèche, library, or music room. Each unit could be added to at need with low-cost, mass-produced modules.

The people living here might be couples, with or without children, unattached adults, grandparents, even orphans, who live close to each other because they want to. The accent would be strongly on flexibility, especially within the modules, which would offer as much or as little space as the occupants wanted or could afford. The location might be an attractive seacoast or rural area, and the town would be one of a number in a cartwheel conformation linked to an industrial, storage and shopping centre.

Because land would be reasonably cheap and available, much of the food used by the community could be intensively 'home-grown', produced either by individuals as a business or cooperatively. Drinking water would come from roofs and wastewater would be recycled into gardens. Much, of not all, of the community's electricity would come from renewable sources, including amorphous photo-voltaic cells on roofs and other surfaces, even on the roads and lanes.

Each 'household', because of its cooperative financing and low land costs, could afford virtually any leisure, sporting or cultural facility, especially to the benefit of the children. These children would have the support of their peer group, as well as access to proxy 'mothers and fathers', and a wide range of educational, sporting and hobby activities. Relationship changes would be cushioned by a supportive group of friends close by. It would be possible for both parents to stay in the same 'house' after they had broken up. The young mother suffering from post-natal depression would be able to find help and support, as well as the possibility of relief from 24-hour-a-day baby care.

Within the perimeter, in which no powered vehicles would be permitted, there would be social and leisure facilities like a swimming pool, also the junior school, located so even small children could find their way to it safely on their own feet. Higher education by expert, gifted communicators would be largely at computer terminals and specialised interactive seminars. All these facilities would be available within walking distance.

Communities a little like this, of course, do exist – there are 256 of them in Israel, with more than 100,000 residents. *Kibbutzim* still produce much of Israel's food, and run the nation's largest network of hotels, holiday villages and country lodgings. A national orchestra and theatre company are based on them. I can recall spending a few days in one – the Leon Blum *kibbutz* not far from the Syrian border. There was an extensive sharing of facilities – most people shared cars rather than owning their own, meals were available in a communal dining room or could be prepared at home.   Orchestras, theatre, a library – all these were fostered by the community as a whole. The place was obviously prosperous, and most people said they liked being there.

Much of the foregoing applies mainly to Western communities, but could be adapted to suit the developing world, where village societies already exist with strong communal tendencies. These could be retained, and built on. Expansion of villages into small industrial towns is possible —it is happening on a large scale in China. Close to a billion people will live in Chinese cities by 2025 – including more new urban dwellers than the entire population of the United States. China plans to build more than 20,000 skyscrapers over the next 20 years, and will provide mass transit services to more than 170 cities. Specializing in innovative industries like nanotechnology, smart materials and advanced pharmaceuticals, these new cities will almost certainly house the largest middle class in the world.

So what about the majority of the human race who live in villages in Asia, Africa and Latin America? Is the answer to move them from these villages into the urban slums? Plainly not, but that is what's happening now. Is it good enough to condemn these people, their children, and their childrens' children to short lifetimes of squalor and hardship? These, after all, are the people of the future. Most of them will be young, while most Westerners will be old. Returning briefly to our main theme, it is dangerous simply to require millions of young people to be under-privileged, without work and with no prospect of a decent life. Studies of civil war, rioting and mob destruction in developing countries has shown again and again that it is mostly young unemployed men who are involved. In South Africa, one of the world's most violent societies, the majority of 46 murders a day are committed by unemployed men under 19. If the Steamroller is the cause of chaos, these young men will be the servants of that chaos. Given a good place to live, and work, their future could be very different.

20 The Trouble with Money

According to an Internet site that envisages world population as 100 people living in a village, six own 59 per cent of all wealth — they are American. Eighty are living in poverty, 50 of these suffering absolute starvation or malnutrition, while 70 are illiterate. Official figures confirm this striking commentary on the global economic situation — on recent UN figures, one per cent of the world's people own 40 per cent of all wealth, while half the planet's population have to share just one per cent of the global pile. Oxfam has concluded that on current trends that wealthy one per cent will own more than half of everything by 2016. According to 2014 figures from London School of Economics researchers Saez and Zucman, 0.01 per cent of Americans, a mere 16000 families, own 22 per cent of that country's wealth, an average of $371 million each. In the 1970s the average American CEO earned 25 times as much as the average worker; by 2014 this had grown to as much as 300 times —some pay envelopes approaching $400million. US gross domestic product growth of 24 per cent since 2000 contrasts with a fall in median (average Joe) pay of 6 per cent over the same period.

Everyone hard up in Britain? Not so. According to the *Sunday Times* Rich List for 2015 Britain's billionaires have doubled their wealth since 2009 to $827billion, while, according to the Equality Trust the wealthiest thousand families have more money than the poorest 40 per cent of Britons, with average households no better off than in 2008. The trust's director, Duncan Exley remarks: 'Inequality at this scale is hugely damaging for society... you're likely to have poor mental health, poorer education, be a victim of violent crime, even die earlier.'

Many of the very rich choose not to use their wealth, often inherited, to any new productive purpose – this financial 'logjam' contributes to a steady decline in growth of the planet's infrastructure, a deficit estimated at as much as a trillion dollars a year. This is a particular problem in India, where, according to the *New Indian Express* 300 'stalled' infrastructure projects are awaiting finance. Other very wealthy people use their money to manipulate currencies, selling massive amounts to drive down the value of money in the victim nation then buying it back cheaply. The man who 'broke the Bank of England', George Soros, made a profit of over a billion dollars trading the British pound in 1992. Such money doesn't come from nowhere, of course — in the end it costs the people of the victim nation.

Gross, and increasing, economic inequality means billions of ordinary people have less money to spend – this was the situation reported almost everywhere in 2014 by retailers whose businesses have suffered from reduced public spending. French economist Thomas Piketty, in his 2014 book *Capital in the 21st Century,* warns that financial inequality will increase, and that this is pretty much automatic in our present financial system. A surprise bestseller, the book seriously questions the values of that system. Perhaps we should note an element of personal risk in unjustifiable economic disparity — in France in 1789 it led to thousands of the rich and privileged losing their heads under the guillotine. While there are no tumbrils rolling through the streets now rich people in many parts of the world are living in gated communities — effectively fortresses — afraid they or their children might be kidnapped. In Brazil's Sao Paulo, one of the world's largest cities, the very rich avoid the streets, flying from place to place in helicopters. Its fleet of 500 choppers is said to be the largest in the world.

None of this is good, worse, it is getting to be dangerous. There seems a plain need for a massive redistribution of wealth, and if we are to believe Piketty, this will not happen unless governments make it happen. A significant wealth tax on incomes over a certain level, with no allowable deductions, would be one way. We could devote the proceeds of that supertax to helping the vast mass of the poor, who would then have money to spend. And the world would be a better place if the tens of billions of tax dollars avoided by most of the world's biggest corporations by 'transferring' income to tax havens could actually be diverted to public exchequers. Less endless talk about this issue and more action would be a very good idea.

I can recall watching images on my TV of enraged crowds in Ireland protesting about the $70 billion bailout of their banks, for which they would ultimately have to pay, one way or another. But this figure was small compared with the hundreds of billions advanced in the

United States, Greece, Iceland – the list goes on and on – causing a resultant burden of debt and privation endured by ordinary people that went on for years. This – the global financial crisis, remember? — was caused by reckless and failed policies indulged in by banks, which, if they had happened in any other area of business, would have resulted in bankruptcy. However the banks, we are told, are too big to fail, they must be 'bailed out'.

Elsewhere we've given some account of the damage done by the big banks' merciless pursuit and manipulation of Third World debt. Now, in spite of the enormity of the GFC bailouts, banks are still paying their people, especially their CEOs, huge salaries and bonuses.

Are private enterprise banks, on their record, the best means available to take care of our money? Money, after all, is important. It is the key to dealing with most of the problems of the planet, the universal solvent, the blood stream of global productive effort. Plainly the way money is used and distributed is basic to the task of eliminating poverty, but most money use seems to increase developing world deprivation rather than otherwise. And there is still that massive burden of unemployment, failed businesses, dwindling demand and savagely depleted incomes in the developed world resulting from the 'global financial crisis.'

There are other kinds of banks. In 1976 a practical idealist – we need lots more of these – started one. He is a Bangladeshi, at that time professor of economics at Chittagong University, who decided to do something about poverty. Muhammad Yusuf founded and

developed the Grameen Bank with the simple initial
gesture of making $27 — yes, twenty seven dollars —
available to 42 hardworking and competent people who
were, in spite of their efforts, living on the edge of
poverty. That poverty, as with that of millions of others,
was due to their oppression by petty but extortionate
moneylenders exacting interest as high as ten per cent a
week. One on the wrong end of this stick was a young
woman supporting three children, who was making
bamboo stools. Yusuf found she depended on a petty
capitalist to supply her with materials, which she could
not afford to buy, on the basis that he bought the finished
products from her at his price. Her 'profit' from
unremitting hard work was the equivalent of two cents a
day, just enough for bare necessities, and small enough
to guarantee her continued poverty – a virtual state of
slavery. Yusuf founded the Grameen Bank to help people
like her — and there are plenty of them.

Conventional bankers wanted nothing to do with
Yusuf's proposition that small amounts of capital made
available to the very poor could transform their lives.
The poor would not repay their loans, they said, the poor
have no collateral. However, by 2014 the Grameen Bank
had over 2000 branches, lending an average of $150 to
more than four million borrowers. Almost all are women,
who repay 98 per cent of the money they borrow, a better
rate than most conventional banks can count on. There is
no written contract between the Grameen Bank and its
borrowers, the outcomes depending purely on trust. The
bank estimates that almost 50 million people have been

lifted from acute poverty because of their loans. Similar 'micro-credit' banks have been started in more than 50 countries.

Using money the micro-credit way plainly has useful and humane results. By contrast conventional money's major influences, calculated interference in national politics and severe distortions in the global economy, are now greater than ever. This is dangerous, partly because it is fundamentally irresponsible. Big business freely acknowledges that its basic concern is not the public welfare, but to increase profits to shareholders. Yet even this limited objective is often betrayed. Concealing or misrepresenting the movement and use of money has become a major industry, resulting in the transfer of millions from shareholders and the public to insider traders.

The collapse of the US energy giant Enron in 2002 prompted this comment: 'Enron hid billions of dollars in debts and operating losses inside private partnerships and dizzyingly complex accounting schemes that were intended to pump up the buzz about the company and support its inflated stock price.' (Time, 28.1.02) Enron executives sold millions of dollars worth of share options they had been granted in the two years before Enron collapsed, while thousands of smaller shareholders lost money as the same shares became almost valueless.

'Greed is good,' as they say on Wall Street. However, manipulating money does not actually create anything. The only true wealth of the world comes from someone somewhere making, creating, or growing something of value. But in today's world the creators get less and less of the retail price of their goods, while the manipulators and middlemen get more and more. The farmer, the writer, the manufacturer, are kept poor, destroying incentive and causing social and economic distortions at a time when innovation and rapid material development in new directions are becoming very important. As the current bias towards the finance, insurance, distribution and service areas continues, with vast accretion of 'paper' wealth, this deficit of useful productivity, real wealth, is becoming painfully obvious.

The new society we need to play our hand efficiently must be flexible and innovative, and be closely geared to rapid and constructive change. Big business, however, tends to resist change. It has investments it wants to be profitable far into the future. The huge sums paid by the energy giants to discredit climate change science and to discourage the development of alternative energy, are typical of this 'conservative' attitude. Every day they can delay an effective transition to sustainable energy is worth millions to them, so it pays to muddy the water.

Misuse of money is ruining a whole continent, Africa, according to Tom Burgis' 2015 book *The Looting Machine.* Burgis, who was an African correspondent for the *Financial Times,* describes a pervasive pattern of manipulation by some of the world's biggest companies, who connive with unscrupulous national leaders to siphon off the continent's huge natural wealth, while tens of millions languish in some of the world's most extreme poverty. Most African nations suffer from this 'resource curse' – they have huge endowments of natural wealth. Zimbabwe has diamonds, there is oil in Angola, uranium in Niger, iron ore in Guinea. Enough money should come from these to make everyone rich, but instead it goes to just a few people. In some cases George Orwell's satire *Animal Farm* approaches reality, with elites who started by sponsoring 'freedom' behaving little differently from the colonial rulers who preceded them.

Then there is 'quantitative easing,' which means creating funny money out of nowhere, basically it has the same effects as printing huge volumes of banknotes. The more of these you pump into economies where productivity is relatively stable, the less the value of the individual unit of money becomes. The idea is to flood banks with capital in the hope of stimulating lending and liquidity. People who are gratified that the house they bought 40 years ago for $20,000 is now worth a million are kidding themselves. In real terms the house is essentially worth pretty much what they paid for it — most of that apparent bonus comes from progressive losses in the value of money.

Naturally, governments and economists don't like to use the term 'funny money', so they came up with 'fiat money', a term significantly obscure to deceive most people – 'fiat' is Latin for 'it shall be'. This is currency a government has made and declared to be legal tender, but which is not backed by a physical commodity, such as gold, and hence has no intrinsic value. Unrestrained printing of fiat money has led to the spectacular collapse of a number of economies — during Germany's experience with it people eventually used bundles of paper money to fuel their furnaces.

There is an awful lot of this funny money around these days. Over a period of five years to 2014 the United States churned out more than $4 trillion of it, equivalent to the total annual income of 20 million American families. (A trillion is one thousand billion, the world's economies measured as GDP amount to about 74 of them). Then Japan got into the act, claiming that cranking the money machine would solve that country's financial problems – incidentally this seems not to have happened. Early in 2015 Europe committed to quantitative easing of more than a trillion euros. This flood of 'new' money has been harmful to most people rather than otherwise, driving up the value of shares and real estate to bizarre levels, and making rich people richer.

The delivery of most of the Greek population into a state of penury is significant because this might also happen to people in other precarious economies. The current predicament of Greece might be compared to that of a fifteen year old to whom a bank has given a very large credit card max. After spending hugely with such apparently easy money, he soon finds difficulty even in meeting the interest payment, much less pay off the principal, and is eventually driven into borrowing more and more. The simple fact that a group of nations has done this to one of their number does not stand close examination — even though the Greeks didn't have to borrow all that money and indulge themselves with large pensions and an early retirement age, and they might have been better about paying their taxes. Granted all that, the money system has condemned millions of people to a severe 'austerity' under which they are suffering acutely. Could a really efficient money system not have done better than this?

Why does the current huge disparity in wealth exist? One major reason is that the rich command big investments in increasingly automated manufacture, while the poor, by and large, are employees, or peasants who labour in their fields using hand tools. Unless there is some sort of control – and there is not — automation means fewer jobs. On the credit side, why should people do boring and repetitive work when automatic machinery and robots can do it quicker and better? Somehow we have to balance these two things, so the undoubted benefits from automation are spread evenly through the community rather than going to a few wealthy corporations. This will not happen by itself, but will demand a considerable degree of government intervention and control. This issue rose during the first decades of the industrial revolution in the 19th century, and resulted in major riots by underpaid workers during which newly built mills were sabotaged and machinery smashed. This movement was called Luddite. If we don't learn from it neo-Luddite responses will happen again.

The increasing number and variety of industrial,

mining and agricultural robots suggests that if their benefits were distributed fairly, humankind could have almost everything it needs at very low cost. Globalization is often criticized, but it can offer a potentially good effect in which the economies of scale could operate to the enormous benefit of humankind, mass-producing useful commodities at little more than the cost of the raw materials. Think watches, ballpoint pens. Yet as automatic processes do more and more work, people will be needed less and less. The workplace – the job – will increasingly fail in its traditional role of distributing wealth and status. Unless these can be provided in other ways the increasing strains of the future will create severe structural weaknesses in society.

Radical changes in the nature of work began during the last decades of Millennium Two. Higher qualifications were required, demand for workers shifted from the youngest and the oldest to the middle-aged, work opportunities for the 'unskilled' dropped, and staff numbers were cut in almost every industry in most parts of the world. 'Downsizing' will continue. One of its worst effects has been the much higher rate of unemployment among young people. Nevertheless the volume of repetitive 'unskilled' work that is now automated is just a beginning, with most of the impact of automation yet to come. Even in the US and Japan, where the use of industrial robots is most advanced, they amount to perhaps ten per cent of the human workforce. Their use is increasing at about six per cent a year, although this figure is growing—according to the think-tank RobotEnomics within two or three decades robots will replace 80 per cent of the current human workforce, who will become, in the words of author, professor and researcher Tyler Cowen, 'a permanent underclass'. Health care and food industry workers are most at risk.

The financial screw-up is getting so bad immediate and effective action is needed to fix it. This will need concerted action by all major nations, not just some. What about the following, for a start:

*A turnover tax, without exemptions, on multi-nationals now avoiding income tax, on their income derived in the taxing country.

*A wealth tax on all incomes over $1million a year.

*A small negative interest rate on money not invested in a productive enterprise.

*Universal wage bonuses for employees, proportionate to company productivity as expressed by profits.

*A stable world currency immune from speculation.

21: Getting Around

Being able to drive to the shops and find what you want there has been taken for granted for so long it may be hard to imagine this not happening. But the world's supply of oil-based fuels has to come to an end some time, and when it does we would be wise to have forms of transport that are not dependent on them. And this crisis could be upon us sooner than we think. Because so much oil is located in trouble-prone parts of the world sudden and extreme failures in supply could happen anytime. In some of the world's most dependent countries, like Australia, with a 91 per cent reliance on imported fuels, petrol station tanks would be empty in a matter of days, and as supply trucks stopped running chilled and frozen food in supermarkets and drugs in pharmacies would run out in a week. We are so dependent on oil that in the event of a war we could be brought to our knees simply by the sinking of a few tankers.

And even without a crisis things will not get better as the years pass. Imagine you are Rip van Winkle. You nod off, feeling unusually tired and don't wake up for 20 years. Having done so, you recall the car needs fuel, so you go off to fill up.

Three hundred pounds! the man wants.

'But...'

'Yeah, three hundred and seven actually... and where's the coupon?'

'Coupon?'

'The ration ticket… Haven't woke up yet?'

This might be just a bad dream, but maybe it's not. Regardless of the brief reprieve offered by coal-seam gas, it's a better than even bet that within two decades motor fuels will at least double in price and be so scarce they'll be rationed almost everywhere. Driving home after paying that bill you reflect, why did I listen to those guys saying 'They'll always find more,' when in fact they are not, nowhere near enough to keep up with future demand.

Oil accounted for a third of all energy use in 2013, consumption increasing to 91 million barrels a day, or 34 billion a year. According to the International Energy Agency daily demand will go up to 115 million barrels by 2035 – if there is that much oil left. On that rate of usage the two largest new oil discoveries over the past 30 years might be enough to meet world demand for not much more than a year. The biggest, off the coast of Brazil, is expected to produce about 20 billion barrels, world use for around eight months. But it will be years before it comes into production and the oil will be very expensive, because it is 160 miles offshore and sitting under four miles of sea, rock and a gigantic layer of salt. Drilling a single exploratory well through this cost $240 million. Kashagan field in the Caspian Sea is regarded as very technically demanding. It is estimated at 13 to 20 billion barrels but it is being delayed by cost over-runs and disputes between the Kazakhstan government and the oil companies.

We are currently still filling our tanks from four 'supergiant' fields discovered decades ago. They are steadily declining, nothing like them has been found since, and careful geophysical surveying indicates no more exist. The Saudi Arabian government owns by far the biggest, Ghawar, which had a peak output around 5 million barrels per day (bpd) in 2009. While, according to one commentator (M. Siddons, *Twilight in the Desert* 2005) the Saudi fields 'could soon approach a serious irreversible decline' they produced a record 10.3 million bpd in 2015, with the International Energy Agency assessing proven reserves at 268 billion barrels, current global oil use for about eight years. Most of this oil is easily recoverable low production cost light crude.

There are four other 'super-giants' – fields that have produced more than a million bpd. Kirkuk, in Iraq, is plagued by continuing warfare in that country, holding production down to around 250,000 bpd. However, in 2015 the Iraqi government still controlled the southern oilfields, which run second to Saudi Arabia with exports of 2.5million bpd. Canterell, in Mexico, peaked at 2.1 million barrels in 2004, but by 2015 was down to less than a quarter of that – barely 325,000. According to the Kuwait Oil Company Burgan, in Kuwait, once the world's second largest oilfield, has dropped from about 2 million bpd to 1.7million. Daqing in China is also definitely in decline, but since independent auditing of reserves is discouraged, the extent of this is largely guesswork. However, in 2015 China was the world's biggest oil importer, buying in 7million bpd, much of which is being stored in reserves.

There is a consensus estimate that world production is dropping at least five per cent a year. Canadian tar-sands and oil shale are available, but are expensive and pollutant, and could produce nowhere near enough output to replace natural oil. Coal-seam gas now being produced as an alternative fuel is still a temporary, finite resource.

And of course, it's not just cars and trucks. Ships, trains use oil, jet aircraft kerosene. So what we face some time in the future is a decline of all our major transport forms, and a drop-off in world trade and tourism more damaging than anything we have ever experienced in peacetime. Supply of food to cities is especially vulnerable – a crisis here could come on quite suddenly. As soon as governments realise an oil drought is imminent they will start commandeering supplies for 'strategic' purposes. Hence there is a need for other ways of moving things and people around, and since new technologies take a long time to come on stream, it would make sense to start developing them now. Countries with natural gas reserves could be well advised to hang on to them, and convert as many of their trucks as possible to compressed natural gas.

Cars — the extent of our love affair with these is indicated by their number, over a billion worldwide, so many it has been surmised an alien traveller looking down from space might believe they're the dominant life form. And they almost all run on petrol and diesel – both derived from oil. However, most of the major carmakers are developing alternative technologies for vehicles that do not rely on oil. These are electric and hydrogen fuel cell cars – neither of which yet provide the range, flexibility and relatively low purchase cost your favourite wagon does.

There is an electric variant, the hybrid, which uses two motors, one petrol, one electric, but since the petrol motor charges the vehicle's batteries as well as propelling it, conventional hybrids still rely on oil-based fuels to cover any distance. I drive a 'plug-in' Chevrolet Volt, which goes around 50 miles on an overnight charge. If you need to go further, a petrol engine cuts in seamlessly, driving a generator that keeps the electric motor going for as long as you put petrol in. Now nearly every major car maker is developing 'plug-in hybrids', which are capable of travelling anything from 50 to 200 miles on battery power alone — the farther it can go, the more the car costs. This is because the lithium-ion batteries all these vehicles use are very expensive.

Fully electric cars like the American Tesla, which has a range of about 300 miles, cost over $100,000. China has the BYD E6 sedan, which has been substantially backed financially by American billionaire Warren Buffet. It will travel 186 miles on one charge, costs about $45,000 and had sold 5000 cars by 2015, mainly used as taxis in China. In 2010 BYD entered a partnership with German carmaker Daimler to produce the Denza, a high quality electric car using lithium iron phosphate batteries. It has a range of 186 miles. With

Mercedes Benz quality and ultra-modern styling, it came on the market in 2015 at around $60,000, selling exclusively in China. Electric cars will come into their own when every petrol station also has quick charging points — these are relatively inexpensive.

Ten years ago the hydrogen fuel cell car was touted as the vehicle of the future, but for a variety of reasons it has failed to materialise in quantity. You can split water into its constituent gases, hydrogen and oxygen, with an electric current. The opposite happens when you combine these two gases – electric power is created, as well as water, which comes out of the exhaust pipe. Take the electricity to a car and you can drive the motor with it. The oxygen comes from the air; hydrogen is in a tank in the car. This seems an elegant technology, but it is complicated, and has other drawbacks. The fuel cells are mildly unreliable and quite expensive, because the platinum catalyst they use costs more than gold.

Storing hydrogen is difficult – the gas has to be compressed or liquefied. Completely new fuelling stations would be needed and this would cost $55 billion in the United States alone. (US National Research Council figure) At present there are only 150 hydrogen fuelling stations in the world. Nevertheless Mercedes is persisting with its F cell car – to the tune of $3 billion in development costs over the last 15 years. After two years of road testing in Iceland, it commenced production of a hydrogen version of its B class – the first fuel cell car in the world to be made for the open market. With advanced technology and a range of 250 miles, it is being sold first in Germany. Hyundai also markets a fuel cell car, and is developing a solar powered refuelling system.

Fast trains are expensive, but they are economic just because they are fast, moving much more freight and many more people along a route than is otherwise possible. In 1993 trains in China averaged 30 miles an hour. By 2007 trains running on 3,700 miles of track were capable of 130 miles an hour. Fast trains are being accepted with enthusiasm by travellers, but are bad news for the aviation industry, especially aircraft flying relatively short inter-city services. Air services are still, to an extent, influenced by weather conditions and can be grounded by such natural events as volcanic ash. They require large stretches of land on which to land and take off, they inflict noise and chemical pollution on communities they fly over, and are notoriously subject to hijacking and terrorist attacks. They need kerosene, which comes from oil, and can operate economically on no other fuel, although bio-diesel can be used as an additive. Such things as fuel cells or solar engines are not practicable except for small low-load glider-like machines. Synthetic kerosene is already being made from coal or natural gas, but these processes are expensive, energy-hungry and polluting, and the raw materials are finite in any case.

How about nuclear aircraft? It's been tried, and it didn't work. The US government put several billion dollars and 15 years work into an attempt to build a nuclear-powered aircraft before the venture was cancelled in 1961 as being too hard. In 1950 the Soviet

Union proposed a nuclear flying boat to weigh a thousand tons, carry a thousand passengers, and fly at a thousand kilometres an hour. This project was also never realised. The United States is looking at nuclear power for its nasty little military drones, the use of which in Afghanistan and Pakistan has killed quite a few children, given the Americans a bad press and probably turned out to be an excellent recruiting tool for the Taliban.

Meanwhile the conventional aircraft seems overdue to a rethink. Progress in recent years has been to make passenger jets bigger, using every possible inch of space to fit in more passengers. Is even more extreme technology the answer? There has been research into 'saddle' type seating that almost has people sitting in each other's laps. Faster aircraft? – the HyperSoar concept is for a passenger aircraft designed to skip in and out of the atmosphere and fly in space at 6,000 miles an hour. Visualised as dart-shaped, rather like a folded paper plane, it is claimed this aircraft could reach any destination on earth from any other point in about two hours and use much less fuel. Passengers might not, however, like the high G rate acceleration and exposure to cancer-causing cosmic radiation. First publicized in 1998, little has been heard of the project since, and nothing in recent years.

Airships, regarded as mildly unreliable and uneconomic in the past, are now back on the agenda. Being lighter than air they use much less fuel – it is even possible they could be clad with solar cells driving electric motors. Cargo blimps powered this way would be slow, but they could be crewless, controlled from electronic waypoints along the route. Electricity might even be beamed up, using microwave links. But what seemed to be intractable problems, especially their need for special handling with a lot of manpower at take-off and landing, have restricted airship development.

HAV304, the world's longest aircraft, is likely to change all that. A hybrid –part airship, part plane – it gets up to 40 per cent of its 'lift' from its hull shape, very much like an aircraft wing. Because of its slight negative buoyancy it can land anywhere, on land or water, with no need for a landing strip. Built by Hybrid Air Vehicles at Cardington, in England, it has been assessed as 70 per cent more environmentally friendly than a conventional aircraft, can stay in the air for three weeks, and cruises at 92 mph. An even bigger version, the Airlander 50, is planned.

Airships can be used for special purposes, like military surveillance. Lockheed Martin has a programme for high altitude blimps, which could be solar powered, and which could remain airborne for as much as a year. Turtle Airships Company claimed in 2007 it was building a prototype for a large, rigid-shelled solar-powered airship planned to be capable of 200 mph, which they would fly around the world in 2011. However, nothing much seems to have happened, other than blogs extolling the virtues of a so far non-existent craft, complete with dance floors and 'fine dining saloons.'

Russia persevered for years with ground-effect

vehicles, which are a cross between an aircraft and a ship, but never got much beyond the prototype phase, and Boeing pecked at the technology for a while. Ground-effect vehicles skim over a land or sea surface only a few feet up, faster than a ship while using much less fuel than an aircraft. Wingship Technology, based in South Korea, has taken up the idea, testing a 50 passenger prototype in 2014. This craft has a delta-shaped wing, allowing it to fly perhaps 20 feet up, so it can handle journeys over moderate waves. It can travel 100 mph for a range of around 600 miles.

Neither electricity nor fuel cells have the potential to replace oil-derived kerosene to power big airliners. A 200 seat aircraft weighs around 115 tonnes at take-off. About a third of this —around 40 tons – is fuel. On one estimate, it would take 3000 tons of lithium-ion batteries to replace this – no airliner could carry this much weight, much less take off with it. Fuel cells are only slightly more feasible. Both Airbus and Boeing are tinkering with the idea of kerosene/electric hybrids, but this is very much at the theoretical stage, and would not solve the basic fuel problem in any case.

To summarise, air travel has a difficult future, except for limited and specialized purposes. There are good, and intractable, reasons for this. Because much of the fuel used in a jetliner is consumed gaining altitude after take-off, short journeys are the most uneconomic, using the most fuel and creating the most pollution. This means inter-city air routes, typically the world's busiest, are open to competition.

Fast trains offer the most feasible alternative, since the often lengthy and tiresome journey to airports is eliminated, and trains able to operate at more than 250 mph can get you there faster and cheaper. Visitors to Shanghai, in China, can travel in a train with a top speed of 300mph, which takes you 18 miles from the airport to the new industrial precinct of Pudong in eight minutes. These are 'maglev 'trains – maglev is short for 'magnetic levitation'. The train literally 'flies' through the air, held a small distance away from its rail or track by powerful magnets, and is propelled by linear electric motors along the track, outside the train itself. This technology is also used for 'pod' vehicles, small two person capsules with 'stations' on sidings off the main line, which is an overhead track. Israel has built one of these systems in Tel Aviv.

The Chinese are now able to run trains on upgraded conventional rail lines at 217mph. These will use a planned network of 16,000 miles of track, at a total cost of about $300 billion, due for completion by 2020. Britain is considering a fast rail link between London, Birmingham, Manchester and Leeds, while US President Obama has allocated an initial $8 billion towards fast rail in that country. Fast trains are operating in France and Germany. However, these represent only a tiny fraction of overall inter-city travel in the world. China seems to be the only nation to recognize the threat posed by the end of oil, and to take early action to deal with it.

Travel between most countries is by air or sea. If long distance air travel is restricted, can we rely on ships to fill the gap? Ships, after all, burn a low grade, highly pollutant oil called bunker fuel, which will also become scarce and expensive as oil reserves are depleted. The answer, according to some people, is to have nuclear ships. Should we? The technology exists, and could certainly be extended, but at the present rate of shipping accidents nuclear propulsion on a large scale would threaten ocean pollution lasting for thousands of years, as well as extreme risks to port cities. Considering the rate of merchant ship losses in both world wars – 3000 Allied ships were sunk in 1941 and 1942 alone – a fleet of nuclear freighters would pose appalling risks if war broke out.

Nuclear power is already used for more than 100 naval ships, mostly submarines and aircraft carriers. Russia has eight nuclear icebreakers, the first of which, *Lenin,* was operational for more than 30 years before she was decommissioned in 1989. Three nuclear-powered merchant ships were built, the US *Savannah,* the German *Otto Hahn,* and the Japanese *Mutsu,* but all were considered uneconomic in the circumstances of the time.

However, a bill was passed in the US House of Representatives in 2008 laying the groundwork for more American naval ships to use nuclear power – it is even a possibility for amphibious assault ships. In order to stay at sea as long as possible their fuel could be enriched to as much as 90 per cent, nuclear weapons grade – this has prompted security concerns. Speakers in the debate quoted a 2007 study by the navy indicating that break-even costs for nuclear propulsion were the equivalent of an oil price of $178 a barrel. If this figure is anywhere near accurate, we are not likely to have nuclear freighters soon.

Studies of long voyages and unreliable journey times in the age of sail indicate that headwinds or calms over short sections of the journey, such as in the tropical doldrums, were the major delaying factors. The availability of cheap fuel – coal, then oil – the short useful life of sailcloth, the brutal conditions sailors worked under, and unreliable weather forecasting all contributed to the demise of commercial sailing ships. Now fuel is getting dearer, durable synthetics are available for sails and weather forecasting is much improved. However the doldrums still exist and winds can be fickle anywhere in the world, so if there is a return to sail the new ships will need to be hybrids, carrying some kind of auxiliary power that does not use oil. This might well be electric, using solar cells and free-wheeling propellers deployed while the ship is under sail, much as regenerative braking is used in hybrid cars.

Since sailing ships are capable of respectable speeds and a considerable cargo capacity, they are quite likely to come back into use, although traditional windjammers won't reappear in large numbers, nor will crews have to clamber aloft to deal with huge square sails. Romantic though these may seem, they were clumsy, dangerous to crews and inefficient. It is significant that modern container ships are now being operated at speeds as low as 12 knots, (14mph) to conserve fuel. The best sailing ships, the tea clippers, were considerably faster than this.

There are practical trading ships now that rely entirely on the sun and the wind for power, although they are based on a technology at least 800 years old. These are the south Pacific *vakas* operating inter-island trade in places like Fiji and the Cook Islands. They are basically double-canoes, a catamaran-like conformation exactly like the 14th century ships that brought the Maoris over thousands of miles of rough sea to their new home, Aotearoa (New Zealand). Modern sailing *vakas* have fibreglass hulls and electric auxiliary motors using solar power. Tough and reliable, they regularly make between 100 and 150 miles a day under sail, even in the rough conditions of the Tasman Sea. Being double-hulled and shallow draft they don't need wharves or jetties to unload cargo and passengers. They are simply run up on to a beach, floating off again on the next high tide.

High-flying kites – SkySails – are already in use as fuel savers in several ships when they are travelling in much the same direction as the wind. Then there are designs around visualising schooner-like ships with four or five masts each carrying identical fore and aft rig. These sails would be raised, lowered and trimmed mechanically, could be readily controlled from a single computer on the bridge, and would be furled or reefed by retracting into the masts. It has even been proposed that computers could constantly assess the weather, and trim sails automatically.

Many of these features exist in a large yacht, the *Maltese Falcon,* the first ship to demonstrate in practical form that sail could be used again for some commercial vessels. Almost 300 feet long, she reached speeds up to 24 knots on her maiden voyage across the Atlantic. *Maltese Falcon* uses something called DynaRig, in which sails are fitted to three rotating masts, which also incorporate the square yards that carry the sails. These are so thoroughly automated that one person can control the ship from a master computer.

Flettner rotors have captured the public imagination for the past 80 years, but it was not until 2010 that they came into practical use in *E -ship 1,* a 13,000 ton 400 feet long freighter built by a German wind turbine manufacturer to transport its products. It carried the first nine turbines from Emden to Dublin in August, 2010. The ship has four Flettner rotors, which look like tall skinny cotton reels rising from the deck. When rotated these act as sails, propelling the ship forward. The ship is a hybrid, using diesel engines as part of its motive force. The rotors are said to allow fuel savings of 30 to 40 per cent at a speed of 16 knots. If this is proved in practice, the technology will have obvious future applications.

All promising, you might say – but it isn't quite as good as it might seem at first. The new technologies are certainly developing, but getting them into general use is another matter. How much time and money would it take replace a billion motor vehicles in the world with electric or fuel cell cars and trucks?

22 Depression and Happiness

Dealing with the challenge of the hazards will need a lot of well-adjusted, productive and determined people — happy individuals who are useful and productive, who get things done and earn money. By contrast unhappy ones, especially if they turn criminal, can cost a mint and are frequently disruptive. There is a gap in our collective attention here – the deliberations of governments, mostly about economics, terror and political infighting, are concerned too little with the rapidly deteriorating quality of life for more and more people.

Should they be more concerned? It comes down again to this matter of sound foundations, and a genuine understanding that large units, like society, nations, are made up of individuals. Can we as a community afford the social and economic cost of a depression epidemic of major proportions? According to some research, even mild depression can be a killer – it has established that people with no history of coronary disease are twice as likely to have a heart attack if they are depressed, even if the depression is only mild. However, death from heart disease is just one depression killer – suicide, a disabled immune system, a tendency towards cancer, are part of a World Health Organization prediction that by 2020 depression will be the second largest killer after heart disease.

'Severe depression is ten times more prevalent than it was fifty years ago.

It assaults women twice as often as men, and it now strikes a full decade earlier in life on average than it did a generation ago,' Martin Seligman reports in his 2006 book *Learned Optimism*. A study of 839 Americans over 30 years to 1999 showed that pessimists – identified through a careful selection process over several years – had a 19 per cent higher deathrate.

So somehow we are failing at the individual level. Our education methods routinely produce significant numbers of illiterate, unbalanced and often amoral people, and our medical advances largely benefit affluent Westerners while the major killing diseases are gaining ground worldwide. Our gadgets frequently deliver disappointing results. Our complex technology, its unforgiving speed, the dilemmas of population, war and peace, government, pollution, productivity – all these things call for management by healthy, balanced, self-confident individuals, yet as time goes by we have fewer and fewer such people. Instead we have the depression epidemic. There is plenty of research that links successful creative work and problem solving with individual happiness and welfare. This implies a huge economic benefit for the world if as many people as possible are happy, healthy and well adjusted.

So how do we get happy? If you look up the word in a dictionary you won't get very far – the *Concise Oxford* leaves it as 'lucky, fortunate, and contented with one's lot.' The popular media push personal and immediate sensual and material gratification as a key to happiness, although these things often result in discontent and unhappiness.

A different insight comes from one of the world's smallest and most remote countries, the Himalayan mountain state of Bhutan, where the index of overall prosperity is not gross national product but gross national happiness. The Centre for Bhutan studies, which is devoted to promoting this idea, has produced *A Short Guide to Gross National Happiness.* According to this barely eight per cent of people are deeply happy, and slightly over ten per cent unhappy. The rest are 'in between'. According to the nation's prime minister, 'We have now clearly distinguished happiness from the fleeting pleasurable "feel good" moods so often associated with the term. True abiding happiness cannot exist while others suffer, and comes only from serving others, living in harmony with nature and realizing our innate wisdom.' Bhutan stubbornly defends its traditional way of life, restricts the number of foreigners coming in, will not allow its mountains to be climbed, and intends to maintain 60 per cent of its land under tree cover.

If you were to write a book called 'How to be Happy,' it would most likely fly off the shelves, since happiness is a commodity most people want more than anything else. Drugs, often enough harmful ones, are developed and eagerly used which, it is hoped, will deliver happiness. Psychologists study the good effects of laughter. All these things show that very large numbers of people are not happy, also that they don't understand – and this is well documented – that the elusive happy state is indeed just that, elusive. It is unlikely to come from taking a pill, it can't be bought no matter how much money you spend, it doesn't necessarily follow success or achievement.

If you believe hundreds of popular songs and 'romantic' novels, to say nothing of traditional folk tales, sexual love does it – 'they lived happily ever after.' But is this true? Falling madly in love has been shown to have such a strong correlation with some forms of mental illness that the time-honoured phrase 'love sick' seems justified – Sigmund Freud described falling in love as 'a kind of sickness and craziness.' Research by Donatella Marazziti at the University of Pisa has shown there are similar changes in brain chemistry in the lovesick and in people with obsessive-compulsive disease, in which sufferers feel compelled to go on repeating certain actions, like making sure a door is locked, or washing their hands. In both cases low levels of the brain 'feel good' neurotransmitter serotonin were observed – around 40 per cent less than in normal people. The same Italian subjects observed a year later after they had become sexually accustomed appeared to have recovered. Their serotonin levels were back to normal.

In a world increasingly dominated by the words 'me, mine' and 'my' it is significant that people who make an effort to help others are often happy – those who are not tend to be wrapped up in themselves, too inward-looking. Research in this area suggests that people who are happy don't think about happiness much.

Being happy, they simply take it for granted. Several studies have concluded that altruism and a strong belief in family values and connections induce a lasting sense of life satisfaction. Having strong religious convictions seems to help.

On the other hand, people who gave career and material success a strong priority tended to become less happy as time went on. American social scientist David Myers describes the experience of happy people as an unselfconscious 'flow' state, and warns that a self-preoccupied pursuit of happiness can detract from that. Happy people, he says, are absorbed in a task that challenges them without overwhelming them. (*The Pursuit of Happiness, 1993*)

Are the depressed people, then, constitutionally unhappy in the sense that this has always been their normal state? There is a popular theory that a genetic influence is involved – that you are born with a default point that sets your level of happiness. Those who believe in this theory say that although events, good or bad, may change your level of happiness for a while, it will always return to your 'set point' within two years. Certainly when Harvard University psychologist Daniel Gilbert consulted more than 100 academics, he found that good and bad things that happened to them had no permanent effect on their usual happiness level, which fairly quickly returned to what was normal for them. University of Illinois researcher Edward Diener concluded that for events like being promoted or losing a lover, most of the effects on mood had gone in three months, with not a trace left after six months. Lottery winners were found to be no happier a year after their win.

Some theorists believe that children can learn unhappiness from their parents – if they see their parents in a permanently unhappy state they may conclude this is normal. And because children tend to see everything in terms of themselves they might blame themselves for their parents' unhappiness. Martin Seligman used experiments with dogs to demonstrate what he called 'learned helplessness.' Faced with mild electric shocks which they could not avoid the dogs simply gave up and accepted the pain. Anyone who has persuaded a dog to take a bath will recognize this. Seligman theorises that human depression comes from a confirmed pessimism about life and its events. 'Life inflicts the same setbacks and tragedies on the optimist as on the pessimist, but the optimist weathers them better.... The optimist bounces back from defeat, and, with his life somewhat poorer, he picks up and starts again. The pessimist gives up and falls into depression.' (*Learned Optimism, P209)*

However, Seligman believes it is possible for the pessimist to become an optimist, using methods he has researched. Put briefly, he says: 'The epidemic of depression stems from the much-noted rise in individualism and the decline in commitment to the common good... a society that exalts the individual to the extent that ours now does will be riddled with depression.' Seligman believes this comes from meaninglessness – 'a lack of attachment to something larger than you are.

23: Priorities

If you've stayed with me this far, you'll appreciate a hard truth – that somehow all the major hazards need to be dealt with, and soon. The longer we keep nuclear weapon stockpiles the more dangerous they get – every day that passes raises the chance of a destructive event. Climate change is also very much a race with time. If we control emissions soon, we will get temperature rises that are unpleasant, but to which we can adapt. Let the issue run on without decisive action and we won't – for most of the world there could be no adapting to 4C or higher. So those two matters warrant a high priority. Keeping the world's population down by fighting poverty and ignorance, and planning ahead to deal with sea level rise could be next to demand action.

It is a sobering thought that all four of these major issues are currently *things we are not really doing much about.* However, zetetics demands that if we know what's good for us, we should be coping with all of them – even though this must involve major shifts away from our current priorities for work and money. As has been suggested earlier, the largest and most obvious shift should be away from spending on weapons, agreed on by multilateral agreement, and negotiating an end to regional wars. The imperative to control groups like daesh is nevertheless recognized. The proposed world police could specialize in controlling such criminal outbreaks, and would probably be a lot more effective than patched up coalitions of national armed forces.

More huge economies could come from eliminating unnecessary manufacturing and packaging. Providing everyone in the world with an electric car and the solar arrays to charge it would do a lot to solve the climate and pollution problems, but perhaps we should go further. The car could be designed to last for many years, perhaps even indefinitely. After all, there are cars 60 years old still on the roads in Cuba, using spare parts made in backyard workshops.

This principle – production of very high quality consumer goods – could be extended to all areas of manufacture. If they were hired out by the year rather than sold there would be an immediate incentive for the maker to design his widget for long life and reliability. Planned obsolescence and unnecessary packaging could then be treated as the crimes they actually are. These thoughts are not put forward from reformist zeal – rather as examples of the fundamental extent of change that is going to be necessary if we want to dodge the Steamroller. Certainly we can't go on the way we are, just playing around at the edges….

Now just a minute… this isn't a pretty picture you're painting for us, some kind of Orwellian dictatorship that'll try and kick us into all sorts of things we don't want… I happen to like Widget cars, and I intend to go on driving Widgets…

I doubt any kind of dictatorship could do that. The necessary discipline and self-sacrifice could, and must, only come from some form of citizen consensus, of the kind that keeps nations going in wartime. Make this big enough and governments will fall in behind you. The people who make up those governments are individuals with children and grandchildren, and in general they are far from stupid. Their cooperation may come more easily than you think.

To get back to priorities – energy and shelter could reasonably rank next. Already the sea is taking hundreds of houses, and the more destructive weather of recent years has destroyed thousands more – consider the millions still cowering in makeshift tents, the huge toll of death and misery caused already by regular flooding and typhoons of unprecedented ferocity, the deaths of so many in Nepal in 2015 as their charming but rickety houses were destroyed by two earthquakes. Humanity as a whole needs to be better protected from such things. More energy efficient, appropriate and much stronger housing will be essential in the dangerous climate we must now expect.

The necessary adaptation and new construction in this area alone will demand energy – lots of it – in circumstances where there will be less and less if we don't take that issue seriously and soon. Building new sustainable energy infrastructure takes energy, and if we do nothing until our current resources run out we will be in big trouble indeed.

You've been reading one person's take so far,

even though it's a take that's used a lot of other peoples' ideas. You might not agree with some things in this book, maybe there are a few you think are plain wrong. None of that is basically important – the aim has been to give you as clear and truthful an account as possible about what is really happening, so you can pick out areas with which you have some familiarity then think about what might be done about them. Your conclusions could be quite different from mine. So what follows is by no means holy writ, rather a road map, an adaptation outline specialized experts might like to work up from.

The point has probably already struck you that it'd be a tall order to deal with all the big hazards together. So it would be if we mounted a generalized haphazard approach to them. But if there is one thing humans are good at it's specializing, and that's what needs to be done. A dedicated international task force needs to be organized to deal with each individual hazard, directing teams of specialist people. There is nothing new or impossible in this, it's the way we've organized the United Nations with its autonomous agencies.

Many of the hazards are complicated and inter-linked, especially the high priority ones and will take time, as well as a large collective effort, to deal with, but none of them is impossible. Some are like a tangled mass of string that needs unravelling. The people who develop the perfect village fireplace will have contributed in a big way to reducing deaths from cancer, but also to the fraught increase in world population. Really get disarmament going, and you'll have a lot of dissatisfied, unemployed people from the weapons factories and the armed services on your hands. Salvage the world's fisheries and there will be a lot of surplus fishing boats and fishermen. There are answers to all of these issues – – once these associations of cause and consequence are taken carefully into account and the problems within them faced squarely there is nothing that can't be sorted out.

An early requirement will be to constantly nag the powerful to get all this done, because it is major industry and governments that are obstructing necessary change. Governments do listen if enough people make enough noise about something, but these same governments are under constant pressure from powerful business and financial groups who want their interests put first. They regularly give money to political parties and, as the saying goes, there is no such thing as a free lunch. The negativity of these pressure groups can only be balanced by a large, determined volume of public opinion, hence, tedious though it may be, individuals of good intent need to throw their time and effort into causes they judge to be important. If they can do this collectively through Internet lobbying sites like AVAAZ, which at the time of writing had 40million members in 194 countries, so much the better – politicians understand numbers that are big enough to count at the next election.

This political 'softening-up' would need to be followed up by an ongoing campaign to influence public opinion, again using the social media and other community links. Schools are of vital importance here. The young people who pass through them will have to deal with the problems of the future, and there can be no excuse for not informing them fully and truthfully about the issues. Some schools will be doing this, but in world terms most don't, or are not equipped to do so.

Why the Steamroller? There do seem to be advantages in giving difficult and complex issues a label. I have used the analogy right through this text after some thought because it does put that label on our set of hazards. The Steamroller, which still means something to most people even though we don't have them any more, seems as good as any other.

Some readers have asked why I haven't emphasized terrorism as a basic hazard — after all, for years governments and the media have talked about it almost incessantly, using it to seriously limit individual freedom and rights. Terrorism is certainly important, one of the most unpleasant and dangerous manifestations of the Steamroller, but it is a consequence, not a cause. There is nothing new about the ugly violence we are seeing. It has flourished throughout history in times of social dissolution.

Nevertheless, it is important, too, to define terrorists accurately, not as people who differ from the official point of view, but based on what they actually do. Governments everywhere have restricted individual freedoms because they think this is necessary for enhanced 'security'. This restraint needs to be balanced very carefully against the need for individual freedoms in a world where we will need plenty of room to move, both physically and ideologically.

As the Steamroller hazards advance they will reach a stage where the major powers will have to decide whether it is better for them — and the world —- to become self-sufficient heavily armed fortresses or to seriously engage with the problems of the planet as a whole. They will need to consider the hard fact that a time must come when no 'fortress' nation, no matter how wealthy and well-armed, could protect itself against the consequences of, say, a global plague, extreme global warming or major contamination following the use of nuclear weapons. Readjustment at that time could prove expensive, painful and probably unsuccessful. A world pattern of action is needed now.

This suggests it is time for serious discussions between the world's major powers to establish at least a road map for global cooperation. Having arrived at a theoretical consensus, courses of action might become more obvious, as well as practical, sensible ways of achieving them. The oceans are crying out for better treatment, so a world Seaforce could be a good start.

So let us consider this world in which we have come to our senses, and we are coping, somewhat uncomfortably with say, a 3 degree temperature increase. What next? If we have any brains, we will have been chastened, we will understand that most of the hazards of the Steamroller are still there, and that the struggle with them must go on. Climate conditions will remain adverse, the sea will go on rising, probably for hundreds of years, there will be less productive land, less food and many fewer humans. Whatever it takes to promote good health and reasonable living conditions among the survivors of all races should become a high priority – this must be associated with a determined campaign to keep down populations, otherwise the falling death rate would again overwhelm us with people.

If the 'new' world is to be more equal and prosperous, people in the wealthier countries must accept a simpler lifestyle in the interests of helping the poor. We ought, at that stage, to have learned that there can be no 'lifeboats' of prosperity floating on a sea of misery and deprivation.

24 Worst Case

Many thousands, possibly millions of people, are so convinced catastrophe is imminent they have gone into a siege mentality to prepare for it, building bunkers under their houses stocked with enough food and water for years, armed with defensive weapons and in some cases surrounded by high walls and barbed wire. These are the survivalists or 'preppers', for whom guarding against the perils of the future has become a way of life.

There are, of course, some perils, such as solar flares or a comet or meteor strike on the earth. The probability of really dangerous intruders from space – objects over 50 feet in diameter – is once or twice every 1000 years. Strikes by objects more than a mile wide that would endanger global civilization are assessed at once every five hundred thousand years.

However, a *National Geographic* article in 2011 considered what damage might be caused in the modern world if there were a solar flare as intense as that in 1859 —the Carrington event— during which the two polar aurorae could be seen almost everywhere in the world. That was, of course, before electricity, computers and modern communications. But now intense solar surges could destroy hundreds of transformers, disrupting power supplies for months on end, computers, lifts, phones would stop working. Other services, like water supply, which are dependent on computers, would also be knocked out.

The fact that we could do little about major natural catastrophes like these makes it all the more important that we do cope with those we can and must influence. The challenge is to do just that, not hide away from them. Although the future will present risks and hardships, many of these could be avoided by changes in the routines of life, rather than complete abdication from the world. Small and trivial though many of these changes may seem, there could be circumstances where they are vital to survival.

So if you're asking, what about me and my family? There is no question — you, like almost everyone else, will need to change your life considerably, and should start planning for that now. If we get warming of 5 or 6 degrees you'll be lucky just to have survived, and your future, and especially that of your children, will be a continual struggle. However, assuming there is some degree of coming to our senses, so we get the 2.5 to 3 degrees that might be reasonably assumed — no better than that and possibly a little worse – adaptation should be a viable option for many people.

If you've read chapter 16 with any care, you'll understand how your risks increase by living in a city, and, very likely, reduce by moving to the country, where you at least may have the space to grow enough food to keep you alive, and to provide your own water and modest amounts of electric power. In the event of war cities are likely to be more vulnerable than they have ever been. Initial attacks would probably use 'electronic bombs', which, when released in the air, deliver a short but intense electro-magnetic pulse, similar to a very close lightning strike, that would knock out all computers – cars, buses, trains, lifts would not work, power and water would be cut off, there would be no supplies of food going to the shops.

This electronic damage could be so severe it could take months, if not years, to repair. Even small countries already have access to this technology, the weapons are cheap and relatively easy to make – you can find basic designs on the Internet – and could easily fall into the hands of blackmailing terrorist groups. All this is understood and accepted by governments, so why do most permit, even encourage, more and more of their people to live in big cities?

Many nations have too many people now. They don't have much spare space, and in the most crowded of these attrition will inevitably be the result — death rates so high populations will reduce abruptly. However, wherever it is possible responsible governments should encourage decentralization and policies to ensure all land is used to its full sustainable potential — not when things turn really bad, but now. Your government probably won't, so you will need to be personally responsible for yourself and your family's welfare. The following are some thoughts about what you might do – in most cases they add up to a high degree of self-sufficiency.

Ideally, acquire your bit of land as soon as if possible, preferably near a creek or a river. If you are this lucky, consider carefully the sort of house you put on it. It should be on one level, built of strong materials, reinforced concrete or mud brick, with a roof well anchored down and with minimal eaves so it won't become airborne in violent winds. Solar tubes and skylights should be installed to get light into dark rooms. The glass in these must be tough enough to resist heavy hail. The house should be set on a hollow reinforced concrete box rather than a slab, providing a large tank into which rainwater can be directed from the roof. Some of the tank space will be devoted to recycling household water by filtering. This will require a small electric pump and an efficient filter system, and you would best use a good quality soap, not detergents, and as little as is necessary of it. Pumps tend to wear out, a complete spare one would be an idea, if not, buy a replacement impeller, seals and other necessary spare parts.

The roof will be a busy place. It will carry solar panels for electricity and a solar hot water heater. Some of it will have grass growing on it — a good form of insulation. Because roofs get lots of sun, you may want to grow some of your vegetables up there – this could be where kids grow things, maybe a herb garden, so put a railing around it. Yet another water tank will be needed up here with its own rain-gathering roof/awning, so growing things can be watered by gravity feed.

Think carefully about the orientation of the house, do some research on passive solar qualities – such things as the location of doors and windows, the use of deciduous trees to shelter sun exposed walls. However, one solid masonry wall exposed to winter sun is a good idea –facing directly north or south, depending on where you live – to act as a heat bank. It should have clerestory windows above it to release heat in summer. The longer this wall is, the better. Also along the lower part of that same wall, locate a greenhouse made of tough clear plastic, not thin glass that could be destroyed by hail. This should be equipped with raised beds of high quality soil, suitable for intensive cultivation. This complex would extend the growing season of vegetables and fruit, while the 'heat-bank' wall would warm the house on winter nights. Keep hens, and perhaps rabbits, to provide protein, maybe also some draught animals, if you have enough land. Goats can provide milk, will eat almost anything, and are easier to deal with than cows.

If you want to avoid the worst aspects of peasant life your vehicle should be a sturdy electric pickup or utility truck rather than a car, so you can carry produce, fertilizer, soil, rocks, compost, whatever. Bear in mind that electric vehicles normally don't like pulling trailers – it upsets their electronics. Your vehicle will need its own set of solar panels to charge it, on a garage or carport roof, enough to give you, say 30 miles of running a day on average, done in the early morning or evening, so you get as many hours of sun-charging as possible. Other family movements should be walking, or on bikes, which could also be electric.

Most people don't have the least idea how to grow their own food, although oddly enough many feel that if they had to, it would be quite easy. Nothing could be further from the truth. Plants, like all life, are what they eat, so the first requirement is top quality soil. This does not always come naturally, your bit of land may have some of it, or it may not. In any case its quality has to be maintained. The classic and best way to do this is composting, which basically means the recycling of soil nutrients. This is done by making compost heaps or bins, in which any kind of organic matter can be broken down quickly into fertile soil. Much of what you now regard as 'rubbish' needs to go into this compost, as well as cover crops grown for this purpose. There are plenty of books around to tell you more about this.

Your house should be provided with composting loos, now a developed technology, and the 'soil' these produce used in the compost, or directly on gardens. These devices use much less water than 'flushing toilets', which waste huge amounts of nutrients by discharging them into the sea, where they do a lot of damage.

Manure from draught or milking animals can also go into compost, or straight into the soil. Best to have a definite space near the house, and with full sunshine all day, for your vegetable and fruit garden. It takes time to build up good soil, so the sooner you start, the better. There should be a stout fence around that garden, because there are innumerable little four-footed or winged people who will be interested in getting to your produce before you do. Hedges providing dense protection can be grown from cuttings. Dogs can be trained to keep birds away, and will stand watch day after day very patiently with a little encouragement. Bear in mind you need to grow enough surplus food for winter, and have the means to preserve it. Fruit and some vegetables can be bottled, and you should have enough solar capacity to run a small deep freezer. You should own a gun.

Learn to co-operate. Trying to be self-sufficient on your own is not easy; there is safety in numbers. Wherever you may be, there will be other people around who are thinking and living like you — a wonderful opportunity. If society has really flown apart, a tribal phase of life will be the beginning of getting it back together, some time, somehow, and so provide the germ of a renewed civilization. Families could specialize, one group keeping cows or growing wheat, another growing vegetables or keeping bees, someone operating as a doctor, another teaching the kids, a third with mechanical skills and tools. Essential drugs like antibiotics, teaching materials, tools and machinery spare parts would need to be stockpiled. Where decisions have to be made, on something like how to distribute the water in a creek, let that be by amicable discussion and consensus. Conditions for bartering goods and services would need to be discussed and set out clearly and definitely. Law must be established, maintained, and, if necessary, imposed. Get this right, and you might be surprised at what a fulfilled and happy community you turn out to be.

Unfortunately lots of you won't be able to get bits of suitable land, which is going to become scarce and very expensive. The solution for most people must be communal, group villages using single-storied structures with curved roofs to withstand high winds, in which self-contained modular living units can be adapted to house families, couples, or individuals. Location on a flat area of fertile soil, preferably near permanent water, or with a valley that could be dammed, would be desirable.

Recapitulating some of the points made in chapter 19, these structures could be grouped in a wide circle, and together accommodate between 400 to 1000 people. Outside the circle would be a ring road, with vehicle parking, shops and other services, and outside that a large circle of community orchards and gardens. Most horizontal surfaces not growing food would be used to store water or generate electricity, using amorphous solar cells that can be coated on any surface, even roadways.

This 'ring' suburb would be one of perhaps thirty, which together would make up a 'town'. The suburbs would be linked by electric tramways, powered by solar arrays along the track, necessarily running infrequently and rather slow. These small 'cities' would include specialized work places; some could have quite advanced areas of manufacturing. Exports to other 'towns' and imports from them could be the beginning of a revived trading economy — perhaps steam trains could be used again.

As you see, dodging the Steamroller will be a complicated business, involving many things that have to be got right — but that's what life's about, isn't it? And a great deal of it will amount to what individuals, individual families, can do. I began this book by saying

I'm hopeful for the future. I still am. Once we've beaten the Steamroller, the conditions of life we've created doing that could lead on to a better future than we've ever had.

Ragged Edge Books
Orders Through Createspace Direct

Dodging the Steamroller:
https://www.createspace.com/5898290
Death is a Scarlet Poppy:
https://www.createspace.com/5918564
Mindcraft of the Paladin:
https://www.createspace.com/5761950
The Valhalla Covenant:
https://www.createspace.com/5805633
Sex Slave or Siren? :
https://www.createspace.com/4113269
Zetetics and the Art of Identity:
https://www.createspace.com/5803485
Indelible Shrine to Youth:
https://www.createspace.com/5862411

www.ingramcontent.com/pod-product-compliance
Lightning Source LLC
Chambersburg PA
CBHW051441170526
45166CB00001B/73